RIVER OUT OF EDEN

RIVER OUT OF EDEN

A Darwinian View of Life

RICHARD DAWKINS

ILLUSTRATIONS BY LALLA WARD

A Division of HarperCollins*Publishers*

The Science Masters Series is a global publishing venture consisting of original science books written by leading scientists and published by a worldwide team of twenty-six publishers assembled by John Brockman. The series was conceived by Anthony Cheetham of Orion Publishers and John Brockman of Brockman Inc., a New York literary agency, and developed in coordination with BasicBooks.

Published by BasicBooks,
A Division of HarperCollins Publishers, Inc.

Designed by Joan Greenfield

Library of Congress Cataloging-in-Publication Data
Dawkins, Richard, 1941–
River out of Eden : A Darwinian view of life / by Richard Dawkins.
p. cm. — (Science Masters series)
Includes bibliographical references and index.
ISBN 0–465–01606–5
1. Genetics. 2. Evolution. I. Title. II. Series.
QH430.D39 1995 94–37146
575—dc20 CIP

95 96 97 98 ❖/RRD 9 8 7 6 5 4 3 2 1

To the memory of

Henry Colyear Dawkins (1921–1992),

Fellow of St. John's College, Oxford:

a master of the art of making things clear.

And a river went out of Eden to water the garden.

—*Genesis 2:10*

CONTENTS

PREFACE

Nature, it seems, is the popular name
For milliards and milliards and milliards
Of particles playing their infinite game
Of billiards and billiards and billiards.

—Piet Hein

Piet Hein captures the classically pristine world of physics. But when the ricochets of atomic billiards chance to put together an object that has a certain, seemingly innocent property, something momentous happens in the universe. That property is an ability to self-replicate; that is, the object is able to use the surrounding materials to make exact copies of itself, including replicas of such minor flaws in copying as may occasionally arise. What will follow from this singular occurrence, anywhere in the universe, is Darwinian selection and hence the baroque extravaganza that, on this planet, we call life. Never were so many facts explained by so few assumptions. Not only does the Darwinian theory command superabundant power to explain. Its economy in doing so has a sinewy elegance, a poetic beauty that outclasses even the most haunting of the world's origin myths. One of my purposes in writing this book has been to accord due recognition to the inspirational quality of our modern understanding of Darwinian life. There is more poetry in Mitochondrial Eve than in her mythological namesake.

The feature of life that, in David Hume's words, most "ravishes into admiration all men who have ever contemplated it" is the complex detail with which its mechanisms—the mechanisms that Charles Darwin called "organs of extreme perfection and complication"—fulfill an apparent purpose. The other feature of earthly life that impresses us is its luxuriant diversity: as measured by estimates of species numbers, there are some tens of millions of different ways of making a living. Another of my purposes is to convince my readers that "ways of making a living" is synonymous with "ways of passing DNA-coded texts on to the future." My "river" is a river of DNA, flowing and branching through geological time, and the metaphor of steep banks confining each species' genetic games turns out to be a surprisingly powerful and helpful explanatory device.

In one way or another, all my books have been devoted to expounding and exploring the almost limitless power of the Darwinian principle—power unleashed whenever and wherever there is enough time for the consequences of primordial self-replication to unfold. *River Out of Eden* continues this mission and brings to an extraterrestrial climax the story of the repercussions that can ensue when the phenomenon of replicators is injected into the hitherto humble game of atomic billiards.

During the writing of this book I have enjoyed support, encouragement, advice and constructive criticism in varying combinations from Michael Birkett, John Brockman, Steve Davies, Daniel Dennett, John Krebs, Sara Lippincott, Jerry

Lyons, and especially my wife, Lalla Ward, who also did the drawings. Some paragraphs here and there are reworked from articles that have appeared elsewhere. The passages of chapter 1 on digital and analog codes are based on my article in *The Spectator* of June 11, 1994. Chapter 3's account of Dan Nilsson and Susanne Pelger's work on the evolution of the eye is partly taken from my "News and Views" article published in *Nature* on April 21, 1994. I acknowledge the editors of both these journals, who commissioned the articles concerned. Finally, I am grateful to John Brockman and Anthony Cheetham for the original invitation to join The Science Masters Series.

Oxford, 1994

RIVER OUT OF EDEN

CHAPTER 1

THE DIGITAL RIVER

All peoples have epic legends about their tribal ancestors, and these legends often formalize themselves into religious cults. People revere and even worship their ancestors—as well they might, for it is real ancestors, not supernatural gods, that hold the key to understanding life. Of all organisms born, the majority die before they come of age. Of the minority that survive and breed, an even smaller minority will have a descendant alive a thousand generations hence. This tiny minority of a minority, this progenitorial élite, is all that future generations will be able to call ancestral. Ancestors are rare, descendants are common.

All organisms that have ever lived—every animal and plant, all bacteria and all fungi, every creeping thing, and all readers of this book—can look back at their ancestors and make the following proud claim: Not a single one of our ancestors died in infancy. They all reached adulthood, and every single one was capable of finding at least one heterosexual partner and of successfully copulating.* Not a single one

*Strictly speaking, there are exceptions. Some animals, like aphids, reproduce without sex. Techniques such as artificial fertilization make it possible for modern humans to have a child without copulating, and even—since eggs for *in vitro* fertilization could be taken from a female fetus—without reaching adulthood. But for most purposes the force of my point is undiminished.

of our ancestors was felled by an enemy, or by a virus, or by a misjudged footstep on a cliff edge, before bringing at least one child into the world. Thousands of our ancestors' contemporaries failed in all these respects, but not a single solitary one of our ancestors failed in any of them. These statements are blindingly obvious, yet from them much follows: much that is curious and unexpected, much that explains and much that astonishes. All these matters will be the subject of this book.

Since all organisms inherit all their genes from their ancestors, rather than from their ancestors' unsuccessful contemporaries, all organisms tend to possess successful genes. They have what it takes to become ancestors—and that means to survive and reproduce. This is why organisms tend to inherit genes with a propensity to build a well-designed machine—a body that actively works as if it is striving to become an ancestor. That is why birds are so good at flying, fish so good at swimming, monkeys so good at climbing, viruses so good at spreading. That is why we love life and love sex and love children. It is because we all, without a single exception, inherit all our genes from an unbroken line of successful ancestors. The world becomes full of organisms that have what it takes to become ancestors. That, in a sentence, is Darwinism. Of course, Darwin said much more than that, and nowadays there is much more we can say, which is why this book doesn't stop here.

There is a natural, and deeply pernicious, way to misunderstand the previous paragraph. It is tempting to think that when ancestors did successful things, the genes they passed on to their children were, as a result, upgraded relative to the genes they had received from their parents.

Something about their success rubbed off on their genes, and that is why their descendants are so good at flying, swimming, courting. Wrong, utterly wrong! Genes do not improve in the using, they are just passed on, unchanged except for very rare random errors. It is not success that makes good genes. It is good genes that make success, and nothing an individual does during its lifetime has any effect whatever upon its genes. Those individuals born with good genes are the most likely to grow up to become successful ancestors; therefore good genes are more likely than bad to get passed on to the future. Each generation is a filter, a sieve: good genes tend to fall through the sieve into the next generation; bad genes tend to end up in bodies that die young or without reproducing. Bad genes may pass through the sieve for a generation or two, perhaps because they have the luck to share a body with good genes. But you need more than luck to navigate successfully through a thousand sieves in succession, one sieve under the other. After a thousand successive generations, the genes that have made it through are likely to be the good ones.

I said that the genes that survive down the generations will be the ones that have succeeded in making ancestors. This is true, but there is one apparent exception I must deal with before the thought of it causes confusion. Some individuals are irrevocably sterile, yet they are seemingly designed to assist the passage of their genes into future generations. Worker ants, bees, wasps and termites are sterile. They labor not to become ancestors but so that their fertile relatives, usually sisters and brothers, will become ancestors. There are two points to understand here. First, in any kind of animal, sisters and brothers have a high probability of sharing copies of the same genes. Second, it is the environment, not the genes, that determines whether an individual termite, say, becomes a reproducer or a sterile worker. All termites contain genes capable of turning them into sterile workers under some environmental conditions, reproducers under other conditions. The reproducers pass on copies of the very same genes that make the sterile workers help them to do so. The sterile workers toil under the influence of genes, copies of which are sitting in the bodies of reproducers. The worker copies of those genes are striving to assist their own reproductive copies through the transgenerational sieve. Termite workers can be male or female; but in ants, bees and wasps the workers are all female; otherwise the principle is the same. In a watered-down form, it also applies to several species of birds, mammals and other animals that exhibit a certain amount of caring for young by elder brothers or sisters. To summarize, genes can buy their way through the sieve, not only by assisting their own body to become an ancestor but by assisting the body of a relation to become an ancestor.

The river of my title is a river of DNA, and it flows through time, not space. It is a river of information, not a river of bones and tissues: a river of abstract instructions for building bodies, not a river of solid bodies themselves. The information passes through bodies and affects them, but it is not affected by them on its way through. The river is not only uninfluenced by the experiences and achievements of the successive bodies through which it flows. It is also uninfluenced by a potential source of contamination that, on the face of it, is much more powerful: sex.

In every one of your cells, half your mother's genes rub

shoulders with half your father's genes. Your maternal genes and your paternal genes conspire with one another most intimately to make you the subtle and indivisible amalgam you are. But the genes themselves do not blend. Only their effects do. The genes themselves have a flintlike integrity. When the time comes to move on to the next generation, a gene either goes into the body of a given child or it does not. Paternal genes and maternal genes do not blend; they recombine independently. A given gene in you came either from your mother or your father. It also came from one, and only one, of your four grandparents; from one, and only one, of your eight great-grandparents; and so on back.

I have spoken of a river of genes, but we could equally well speak of a band of good companions marching through geological time. All the genes of one breeding population are, in the long run, companions of each other. In the short run, they sit in individual bodies and are temporarily more intimate companions of the other genes sharing that body. Genes survive down the ages only if they are good at building bodies that are good at living and reproducing in the particular way of life chosen by the species. But there is more to it than this. To be good at surviving, a gene must be good at working together with the other genes in the same species—the same river. To survive in the long run, a gene must be a good companion. It must do well in the company of, or against the background of, the other genes in the same river. Genes of another species are in a different river. They do not have to get on well together—not in the same sense, anyway—for they do not have to share the same bodies.

The feature that defines a species is that all members of

any one species have the same river of genes flowing through them, and all the genes in a species have to be prepared to be good companions of one another. A new species comes into existence when an existing species divides into two. The river of genes forks in time. From a gene's point of view, *speciation*, the origin of new species, is "the long goodbye." After a brief period of partial separation, the two rivers go their separate ways forever, or until one or the other dries extinct into the sand. Secure within the banks of either river, the water is mixed and remixed by sexual recombination. But water never leaps its banks to contaminate the other river. After a species has divided, the two sets of genes are no longer companions. They no longer meet in the same bodies and they are no longer required to get on well together. There is no longer any intercourse between them—and intercourse here means, literally, sexual intercourse between their temporary vehicles, their bodies.

Why should two species divide? What initiates the long goodbye of their genes? What provokes a river to split and the two branches to drift apart, never to meet again? The details are controversial, but nobody doubts that the most important ingredient is accidental geographical separation. The river of genes flows in time, but the physical repartnering of genes takes place in solid bodies, and bodies occupy a location in space. A gray squirrel in North America would be capable of breeding with a gray squirrel in England, if they ever met. But they are unlikely to meet. The river of gray-squirrel genes in North America is effectively separated, by three thousand miles of ocean, from the river of gray-squirrel genes in England. The two bands of genes are no longer companions in fact, although they are still presumably capable of acting as good companions should the opportunity arise. They have said farewell, though it is not an irrevocable goodbye—yet. But given another few thousand years of separation, it is probable that the two rivers will have drifted so far apart that if individual squirrels meet, they will no longer be able to exchange genes. "Drift apart" here means apart not in space but in compatibility.

Something like this almost certainly lies behind the older separation between gray squirrels and red squirrels. They cannot interbreed. They overlap geographically in parts of Europe and, although they meet and probably confront one another over disputed nuts from time to time, they cannot mate to produce fertile offspring. Their genetic rivers have drifted too far apart, which is to say that their genes are no longer well suited to cooperate with one another in bodies. Many generations ago, ancestors of gray squirrels and ancestors of red squirrels were one and the same individuals. But they became geographically separated—perhaps by a mountain range, perhaps by water, eventually by the Atlantic Ocean. And their genetic ensembles grew apart. Geographical separation bred a lack of compatibility. Good companions became poor companions (or they would turn out to be poor companions if put to the test in a mating encounter). Poor companions became poorer still, until now they are not companions at all. Their goodbye is final. The two rivers are separate and destined to become more and more separate. The same story underlies the much earlier separation between, say, our ancestors and the ancestors of elephants. Or between ostrich ancestors (which were also our ancestors) and the ancestors of scorpions.

There are now perhaps thirty million branches to the river

of DNA, for that is an estimate of the number of species on earth. It has also been estimated that the surviving species constitute about 1 percent of the species that have ever lived. It would follow that there have been some three billion branches to the river of DNA altogether. Today's thirty million branch rivers are irrevocably separate. Many of them are destined to wither into nothing, for most species go extinct. If you follow the thirty million rivers (for brevity, I'll refer to the branch rivers as rivers) back into the past, you will find that, one by one, they join up with other rivers. The river of human genes joins with the river of chimpanzee genes at about the same time as the river of gorilla genes does, some seven million years ago. A few million years farther back, our shared African ape river is joined by the stream of orangutan genes. Farther back still, we are joined by a river of gibbon genes—a river that splits downstream into a number of separate species of gibbon and siamang. As we push on backward in time, our genetic river unites with rivers destined, if followed forward again, to branch into the Old World monkeys, the New World monkeys, and the lemurs of Madagascar. Even farther back, our river unites with those leading to other major groups of mammals: rodents, cats, bats, elephants. After that, we meet the streams leading to various kinds of reptiles, birds, amphibians, fish, invertebrates.

Now here is an important respect in which we have to be cautious about the river metaphor. When we think of the divide leading to all the mammals—as opposed to, say, the stream leading to the gray squirrel—it is tempting to imagine something on a grand, Mississippi/Missouri scale. The mammal branch is, after all, destined to branch and branch and branch again, until it produces all the mammals—from pigmy shrew to elephant, from moles underground to monkeys atop the canopy. The mammal branch of the river is destined to feed so many thousands of important trunk waterways, how could it be other than a massive, rolling torrent? But this image is deeply wrong. When the ancestors of all the modern mammals broke away from those that are not mammals, the event was no more momentous than any other speciation. It would have gone unremarked by any naturalist who happened to be around at the time. The new branch of the river of genes would have been a trickle, inhabiting a species of little nocturnal creature no more different from its nonmammalian cousins than a red squirrel is from a gray. It is only with hindsight that we see the ancestral mammal as a mammal at all. In those days, it would have been just another species of mammal-like reptile, not markedly different from perhaps a dozen other small, snouty, insectivorous morsels of dinosaur food.

The same lack of drama would have attended the earlier splits between the ancestors of all the great groups of animals: the vertebrates, the mollusks, the crustaceans, the insects, the segmented worms, the flatworms, the jellyfish and so on. When the river that was to lead to the mollusks (and others) parted from the river that was to lead to the vertebrates (and others), the two populations of (probably wormlike) creatures would have been so alike that they could have mated with one another. The only reason they didn't is that they had become accidentally separated by some geographical barrier, perhaps dry land separating previously united waters. Nobody could have guessed that one population was destined to spawn the mollusks and the other the vertebrates. The two rivers of DNA were streamlets barely parted, and the two groups of animals were all but indistinguishable.

Zoologists know all this, but they forget it sometimes when contemplating the really big animal groups, like mollusks and vertebrates. They are tempted to think of the divide between major groups as a momentous event. The reason zoologists may be so misled is that they have been brought up in the almost reverential belief that each of the great divisions of the animal kingdom is furnished with something deeply unique, often called by the German word *Bauplan.* Although this word just means "blueprint," it has become a recognized technical term, and I shall inflect it as an English word, even though (as I am slightly shocked to discover) it is not yet in the current edition of the Oxford English Dictionary. (Since I enjoy the word less than some of my colleagues do, I admit to a tiny *frisson* of *Schadenfreude* at its absence; those two foreign words *are* in the Dictionary, so there is no systematic prejudice against importation.) In its technical sense, bauplan is often translated as "fundamental body plan." The use of the word "fundamental" (or, equivalently, the self-conscious dropping into German to indicate profundity) is what causes the damage. It can lead zoologists to make serious errors.

One zoologist, for instance, has suggested that evolution in the Cambrian period (between about six hundred million and about five hundred million years ago) must have been a completely different kind of process from evolution in later times. His reasoning was that nowadays it is new species that are coming into existence, whereas in the Cambrian period major groups were appearing, such as the mollusks and the crustaceans. The fallacy is glaring! Even creatures as radically different from one another as mollusks and crustaceans were originally just geographically separated populations of the same species. For a while, they could have interbred if they had met, but they did not. After millions of years of separate evolution, they acquired the characteristics which we, with the hindsight of modern zoologists, now recognize as those of mollusks and crustaceans respectively. These characteristics are dignified with the grandiose title of "fundamental body plan" or "bauplan." But the major bauplans of the animal kingdom diverged from common origins by gradual degrees.

Admittedly, there is a minor, if much publicized, disagreement over quite *how* gradual or "jumpy" evolution is. But nobody, and I mean nobody, thinks that evolution has ever been jumpy enough to invent a whole new bauplan in one step. The author I quoted was writing in 1958. Few zoologists would explicitly take his position today, but they sometimes do so implicitly, speaking as though the major groups of animals arose spontaneously and perfectly formed, like Athena from the head of Zeus, rather than by divergence of an ancestral population while in accidental geographical isolation.*

The study of molecular biology has, in any case, shown the great animal groups to be much closer to one another than we used to think. You can treat the genetic code as a dictionary in which sixty-four words in one language (the sixty-four

*Readers might like to keep these points in mind when consulting *Wonderful Life,* Stephen J. Gould's beautifully written account of the Burgess Shale Cambrian fauna.

possible triplets of a four-letter alphabet) are mapped onto twenty-one words in another language (twenty amino acids plus a punctuation mark). The odds of arriving at the same 64:21 mapping twice by chance are less than one in a million million million million million. Yet the genetic code is in fact literally identical in all animals, plants and bacteria that have ever been looked at. All earthly living things are certainly descended from a single ancestor. Nobody would dispute that, but some startlingly close resemblances between, for instance, insects and vertebrates are now showing up when people examine not just the code itself but detailed sequences of genetic information. There is a quite complicated genetic mechanism responsible for the segmented body plan of insects. An uncannily similar piece of genetic machinery has also been found in mammals. From a molecular point of view, all animals are pretty close relatives of one another and even of plants. You have to go to bacteria to find our distant cousins, and even then the genetic code itself is identical to ours. The reason it is possible to do such precise calculations on the genetic code but not on the anatomy of bauplans is that the genetic code is strictly digital, and digits are things you can count precisely. The river of genes is a digital river, and I must now explain what this engineering term means.

Engineers make an important distinction between digital and analog codes. Phonographs and tape recorders—and until recently most telephones—use analog codes. Compact disks, computers, and most modern telephone systems use digital codes. In an analog telephone system, continuously fluctuating waves of pressure in the air (sounds) are transduced into correspondingly fluctuating waves of voltage in a wire. A phonograph record works in a similar way: the wavy grooves cause a stylus to vibrate, and the movements of the stylus are transduced into corresponding fluctuations in voltage. At the other end of the line these voltage waves are reconverted, by a vibrating membrane in the telephone's earpiece or the phonograph's loudspeaker, back into the corresponding air-pressure waves, so that we can hear them. The code is a simple and direct one: electrical fluctuations in wire are proportional to pressure fluctuations in air. All possible voltages, within certain limits, may pass down the wire, and the differences between them matter.

In a digital telephone, only two possible voltages—or some other discrete number of possible voltages, such as 8 or 256—pass down the wire. The information lies not in the voltages themselves but in the patterning of the discrete levels. This is called Pulse Code Modulation. The actual voltage at any one time will seldom be exactly equal to any of the eight, say, nominal values, but the receiving apparatus will round it off to the nearest of the designated voltages, so that what emerges at the other end of the line is well-nigh perfect even if the transmission along the line is poor. All you have to do is set the discrete levels far enough apart so that random fluctuations can never be misinterpreted by the receiving instrument as the wrong level. This is the great virtue of digital codes, and it is why audio and video systems—and information technology generally—are increasingly going digital. Computers, of course, use digital codes for everything they do. For reasons of convenience, it is a binary code—that is, it has only two levels of voltage instead of 8 or 256.

Even in a digital telephone, the sounds entering the

mouthpiece and leaving the earpiece are still analog fluctuations in air pressure. It is the information traveling from exchange to exchange that is digital. Some kind of code has to be set up to translate analog values, microsecond by microsecond, into sequences of discrete pulses—digitally coded numbers. When you plead with your lover over the telephone, every nuance, every catch in the voice, every passionate sigh and yearning timbre is carried along the wire solely in the form of numbers. You can be moved to tears by numbers—provided they are encoded and decoded fast enough. Modern electronic switching gear is so fast that the line's time can be divided into slices, rather as a chess master may divide his time among twenty games in rotation. By this means, thousands of conversations can be slotted into the same telephone line, apparently simultaneously yet electronically segregated without interference. A trunk data line—many of them nowadays are not wires at all but radio beams, either transmitted directly from hilltop to hilltop or bounced off satellites—is a massive river of digits. But because of this ingenious electronic segregation, it is thousands of digital rivers, which share the same banks only in a superficial sense—like red squirrels and gray, who share the same trees but never intermingle their genes.

Back in the world of engineers, the deficiencies of analog signals don't matter too much, as long as they aren't copied repeatedly. A tape recording may have so little hiss on it that you hardly notice it—unless you amplify the sound, in which case you amplify the hiss and introduce some new noise too. But if you make a tape of the tape, then a tape of the tape of the tape, and so on and on, after a hundred "generations" a horrible hiss will be all that remains. Something like this was a problem in the days when telephones were all analog. Any telephone signal fades over a long wire and has to be boosted—reamplified—every hundred miles or so. In analog days this was a bugbear, because each amplification stage increased the proportion of background hiss. Digital signals, too, need boosting. But, for the reason we've seen, the boosting does not introduce any error: things can be set up so that the information gets through perfectly, no matter how many boosting stations intervene. Hiss does not increase even over hundreds and hundreds of miles.

When I was a small child, my mother explained to me that our nerve cells are the telephone wires of the body. But are they analog or digital? The answer is that they are an interesting mixture of both. A nerve cell is not like an electric wire. It is a long thin tube along which waves of chemical change pass, like a trail of gunpowder fizzing along the ground—except that, unlike a trail of gunpowder, the nerve soon recovers and can fizz again after a short rest period. The absolute magnitude of the wave—the temperature of the gunpowder—may fluctuate as it races along the nerve, but this is irrelevant. The code ignores it. Either the chemical pulse is there or it is not, like two discrete voltage levels in a digital telephone. To this extent, the nervous system is digital. But nerve impulses are not dragooned into bytes: they don't assemble into discrete code numbers. Instead, the strength of the message (the loudness of the sound, the brightness of the light, maybe even the agony of the emotion) is encoded as the rate of impulses. Engineers know this as Pulse Frequency Modulation, and it was popular with them before Pulse Code Modulation was adopted.

A pulse rate is an analog quantity, but the pulses themselves are digital: they are either there or they are not, with no half measures. And the nervous system reaps the same benefit from this as any digital system does. Because of the way nerve cells work, there is the equivalent of an amplifying booster, not every hundred miles but every millimeter—eight hundred boosting stations between the spinal cord and your fingertip. If the absolute height of the nerve impulse—the gunpowder wave—mattered, the message would be distorted beyond recognition over the length of a human arm, let alone a giraffe's neck. Each stage in the amplification would introduce more random error, like what happens when a tape recording is made of a tape recording eight hundred times over. Or when you Xerox a Xerox of a Xerox. After eight hundred photocopying "generations," all that's left is a gray blur. Digital coding offers the only solution to the nerve cell's problem, and natural selection has duly adopted it. The same is true of genes.

Francis Crick and James Watson, the unravelers of the molecular structure of the gene, should, I believe be honored for as many centuries as Aristotle and Plato. Their Nobel Prizes were awarded "in physiology or medicine," and this is right but almost trivial. To talk of continuous revolution is almost a contradiction in terms, yet not only medicine but our whole understanding of life will go on being revolutionized again and again as a direct result of the change in thinking that those two young men initiated in 1953. Genes themselves, and genetic disease, are only the tip of the iceberg. What is truly revolutionary about molecular biology in the post–Watson-Crick era is that it has become digital.

After Watson and Crick, we know that genes themselves, within their minute internal structure, are long strings of pure digital information. What is more, they are truly digital, in the full and strong sense of computers and compact disks, not in the weak sense of the nervous system. The genetic code is not a binary code as in computers, nor an eight-level code as in some telephone systems, but a quaternary code, with four symbols. The machine code of the genes is uncannily computerlike. Apart from differences in jargon, the pages of a molecular-biology journal might be interchanged with those of a computer-engineering journal. Among many other consequences, this digital revolution at the very core of life has dealt the final, killing blow to vitalism—the belief that living material is deeply distinct from nonliving material. Up until 1953 it was still possible to believe that there was something fundamentally and irreducibly mysterious in living protoplasm. No longer. Even those philosophers who had been predisposed to a mechanistic view of life would not have dared hope for such total fulfillment of their wildest dreams.

The following science-fiction plot is feasible, given a technology that differs from today's only in being a little speeded up. Professor Jim Crickson has been kidnapped by an evil foreign power and forced to work in its biological-warfare labs. To save civilization it is vitally important that he should communicate some top-secret information to the outside world, but all normal channels of communication are denied him. Except one. The DNA code consists of sixty-four triplet "codons," enough for a complete upper- and lower-case English alphabet plus ten numerals, a space character and a full stop. Professor Crickson takes a virulent influenza virus off the laboratory shelf and engineers into its genome the complete text of his message to the outside world, in perfectly formed English sentences. He repeats his message over and over again in the engineered genome, adding an easily recognizable "flag" sequence—say, the first ten prime numbers. He then infects himself with the virus and sneezes in a room full of people. A wave of flu sweeps the world, and medical labs in distant lands set to work to sequence its genome in an attempt to design a vaccine. It soon becomes apparent that there is a strange repeated pattern in the genome. Alerted by the prime numbers—which cannot have arisen spontaneously—somebody tumbles to the idea of deploying code-breaking techniques. From there it would be short work to read the full English text of Professor Crickson's message, sneezed around the world.

Our genetic system, which is the universal system of all life on the planet, is digital to the core. With word-for-word accuracy, you could encode the whole of the New Testament in those parts of the human genome that are at present filled with "junk" DNA—that is, DNA not used, at least in the ordinary way, by the body. Every cell in your body contains the equivalent of forty-six immense data tapes, reeling off digital characters via numerous reading heads working simultaneously. In every cell, these tapes—the chromosomes—contain the same information, but the reading heads in different kinds of cells seek out different parts of the database for their own specialist purposes. That is why muscle cells are different from liver cells. There is no spirit-driven life force, no throbbing, heaving, pullulating, protoplasmic, mystic jelly. Life is just bytes and bytes and bytes of digital information.

Genes are pure information—information that can be encoded, recoded and decoded, without any degradation or change of meaning. Pure information can be copied and, since it is digital information, the fidelity of the copying can be immense. DNA characters are copied with an accuracy that rivals anything modern engineers can do. They are copied down the generations, with just enough occasional errors to introduce variety. Among this variety, those coded combinations that become more numerous in the world will obviously and automatically be the ones that, when decoded and obeyed inside bodies, make those bodies take active steps to preserve and propagate those same DNA messages. We—and that means all living things—are survival machines programmed to propagate the digital database that did the programming. Darwinism is now seen to be the survival of the survivors at the level of pure, digital code.

With hindsight, it could not have been otherwise. An analog genetic system could be imagined. But we have already seen what happens to analog information when it is recopied over successive generations. It is Chinese Whispers. Boosted telephone systems, recopied tapes, photocopies of photocopies—analog signals are so vulnerable to cumulative degradation that copying cannot be sustained beyond a limited number of generations. Genes, on the other hand, can self-copy for ten million generations and scarcely degrade at all. Darwinism works only because—apart from discrete mutations, which natural selection either weeds out or preserves—the copying process is perfect. Only a digital genetic system is capable of sustaining Darwinism over eons of geological time.

Nineteen fifty-three, the year of the double helix, will come to be seen not only as the end of mystical and obscurantist views of life; Darwinians will see it as the year their subject went finally digital.

The river of pure digital information, majestically flowing through geological time and splitting into three billion branches, is a powerful image. But where does it leave the familiar features of life? Where does it leave bodies, hands and feet, eyes and brains and whiskers, leaves and trunks and roots? Where does it leave us and our parts? We—we animals, plants, protozoa, fungi and bacteria—are we just the banks through which rivulets of digital data flow? In one sense, yes. But there is, as I have implied, more to it than that. Genes don't only make copies of themselves, which flow on down the generations. They actually spend their time in bodies, and they influence the shape and behavior of the successive bodies in which they find themselves. Bodies are important too.

The body of, say, a polar bear is not just a pair of riverbanks for a digital streamlet. It is also a machine of bear-sized complexity. All the genes of the whole population of polar bears are a collective—good companions, jostling with one another through time. But they do not spend all the time in the company of all the other members of the collective: they change partners within the set that is the collective. The collective is defined as the set of genes that can potentially meet any other genes in the collective (but no member of any of the thirty million other collectives in the world). The actual meetings always take place inside a cell in a polar bear's body. And that body is not a passive receptacle for DNA.

For a start, the sheer number of cells, in every one of which is a complete set of genes, staggers the imagination: about nine hundred million million for a large male bear. If you lined up all the cells of a single polar bear in a row, the array would comfortably make the round trip from here to the moon and back. These cells are of a couple of hundred distinct types, essentially the same couple of hundred for all mammals: muscle cells, nerve cells, bone cells, skin cells and so on. Cells of any one of these distinct types are massed together to form tissues: muscle tissue, bone tissue and so on. All the different types of cells contain the genetic instructions needed to make any of the types. Only the genes appropriate to the tissue concerned are switched on. This is why cells of the different tissues are of different shapes and sizes. More interestingly, the genes switched on in the cells of a particular type cause those cells to grow their tissues into particular shapes. Bones are not shapeless masses of hard, rigid tissue. Bones have particular shapes, with hollow shafts, balls and sockets, spines and spurs. Cells are programmed, by the genes switched on inside them, to behave as if they know where they are in relation to their neighboring cells, which is how they build their tissues up into the shape of ear lobes and heart valves, eye lenses and sphincter muscles.

The complexity of an organism such as a polar bear is many-layered. The body is a complex collection of precisely shaped organs, like livers and kidneys and bones. Each organ is a complex edifice fashioned from particular tissues whose building bricks are cells, often in layers or sheets but often in solid masses too. On a much smaller scale, each cell has a highly complex interior structure of folded membranes. These membranes, and the water between them, are the scene of intricate chemical reactions of very numerous distinct types. A chemical factory belonging to ICI or Union Carbide may have several hundred distinct chemical reactions going on inside it. These chemical reactions will be kept separate from one another by the walls of the flasks, tubes and so on. A living cell might have a similar number of chemical reactions going on inside it simultaneously. To some extent the membranes in a cell are like the glassware in a laboratory, but the analogy is not a good one for two reasons. First, although many of the chemical reactions go on between the membranes, a good many go on *within* the substance of the membranes themselves. Second, there is a more important way in which the different reactions are kept separate. Each reaction is catalyzed by its own special enzyme.

An enzyme is a very large molecule whose three-dimensional shape speeds up one particular kind of chemical reaction by providing a surface that encourages that reaction. Since what matters about biological molecules is their three-dimensional shape, we could regard an enzyme as a large machine tool, carefully jigged to turn out a production line of molecules of a particular shape. Any one cell, therefore, may have hundreds of separate chemical reactions going on inside it simultaneously and separately, on the surfaces of different enzyme molecules. Which particular chemical reactions go on in a given cell is determined by which particular kinds of enzyme molecules are present in large numbers. Each enzyme molecule, including its all-important shape, is assembled under the deterministic influence of a particular gene. To be specific, the precise sequence of several hundred code letters in the gene determines, by a set of rules that are totally known (the genetic code), the sequence of amino acids in the enzyme molecule. Every enzyme molecule is a linear chain of amino acids, and every linear chain of amino acids spontaneously coils up into a unique and particular three-dimensional structure, like a knot, in which parts of the chain form cross-links with other parts of the chain. The exact three-dimensional structure of the knot is determined by the one-dimensional sequence of amino acids, and therefore by the one-dimensional sequence of code letters in the gene. And thus the chemical reactions that take place in a cell are determined by which genes are switched on.

What, then, determines which genes are switched on in a particular cell? The answer is the chemicals that are already present in the cell. There is an element of chicken-and-egg paradox here, but it is not insuperable. The solution to the paradox is actually very simple in principle, although complicated in detail. It is the solution that computer engineers know as bootstrapping. When I first started using computers, in the 1960s, all programs had to be loaded via paper tape. (American computers of the period often used punched cards, but the principle was the same.) Before you could load in the large tape of a serious program, you had to load in a smaller program called the bootstrap loader. The bootstrap loader was a program to do one thing: to tell the computer how to load paper tapes. But—here is the chicken-and-egg paradox—how was the bootstrap-loader tape itself loaded? In modern computers, the equivalent of the bootstrap loader is hardwired into the machine, but in those early days you had to begin by toggling switches in a ritually patterned sequence. This sequence told the computer how to begin to read the first part of the bootstrap-loader tape. The first part of the bootstrap-loader tape then told it a bit more about how to read the next part of the bootstrap-loader tape and so on. By the time the whole bootstrap loader had been sucked in, the computer knew how to read any paper tape, and it had become a useful computer.

When an embryo begins, a single cell, the fertilized egg, divides into two; each of the two divides into four; each of the four divides to make eight, and so on. It takes only a few dozen generations to work the cell numbers up into the trillions, such is the power of exponential division. But, if this were all there was to it, the trillions of cells would all be the same. How, instead, do they differentiate (to use the technical term) into liver cells, kidney cells, muscle cells and so on, each with different genes turned on and different enzymes active? By bootstrapping, and it works like this. Although the egg looks like a sphere, it actually has polarity in its internal chemistry. It has a top and a bottom and, in many cases, a front and a rear (and therefore also a left and a right side) as well. These polarities show themselves in the form of gradients of chemicals. Concentrations of some chemicals steadily increase as you move from front to rear, others as you move from top to bottom. These early gradients are pretty simple, but they are enough to form the first stage in a bootstrapping operation.

When the egg has divided into, say, thirty-two cells—that is, after five divisions—some of those thirty-two cells will have more than their fair share of topside chemicals, others more than their fair share of bottomside chemicals. The cells may also be unbalanced with respect to the chemicals of the fore-and-aft gradient. These differences are enough to cause different combinations of genes to be turned on in different cells. Therefore different combinations of enzymes will be present in the cells of different parts of the early embryo. This will see to it that different combinations of further genes are turned on in different cells. Lineages of cells diverge, therefore, instead of remaining identical to their clone-ancestors within the embryo.

These divergences are very different from the divergences of species we talked about earlier. These cell divergences are programmed and predictable in detail, whereas those species divergences were the fortuitous results of geographical accidents and were unpredictable. Moreover, when species diverge, the genes themselves diverge, in what I fancifully called the long goodbye. When cell lineages within an embryo diverge, both divisions receive the same genes—all of them. But different cells receive different combinations of chemicals, which switch on different combinations of genes, and some genes work to switch other genes on or off. And so the bootstrapping continues, until we have the full repertoire of different kinds of cells.

The developing embryo doesn't just differentiate into a couple of hundred different types of cells. It also undergoes elegant dynamic changes in external and internal shape. Perhaps the most dramatic of these is one of the earliest: the process known as gastrulation. The distinguished embryologist Lewis Wolpert has gone so far as to say, "It is not birth, marriage, or death, but gastrulation which is truly the most important time in your life." What happens at gastrulation is that a hollow ball of cells buckles to form a cup with an inner lining. Essentially all embryologies throughout the animal kingdom undergo this same process of gastrulation. It is the uniform foundation on which the diversity of embryologies rests. Here I mention gastrulation as just one example—a particularly dramatic one—of the kind of restless, origami-like movement of whole sheets of cells that is often seen in embryonic development.

At the end of a virtuoso origami performance; after numerous foldings-in, pushings-out, bulgings and stretchings of layers of cells; after much dynamically orchestrated differential growth of parts of the embryo at the expense of other parts; after differentiation into hundreds of chemically and physically specialized kinds of cells; when the total number of cells has reached into the trillions, the final product is a baby. No, even the baby is not final, because the whole growth of the individual—again, with some parts growing faster than others—past adulthood into old age should be seen as an extension of the same process of embryology: total embryology.

Individuals vary because of differences in quantitative details in their total embryology. A layer of cells grows a little farther before folding in on itself, and the result is—What?—an aquiline rather than a retroussé nose; flat feet, which may save your life because they debar you from the Army; a particular conformation of the shoulder blade that predisposes you to be good at throwing spears (or hand grenades, or cricket balls, depending on your circumstances). Sometimes individual changes in the origami of cell layers can have tragic consequences, as when a baby is born with stumps for arms and no hands. Individual differences that do not manifest themselves in cell-layer origami but purely chemically may be no less important in their consequences: an inability to digest milk, a predisposition to homosexuality, or to peanut allergy, or to think that mangos taste offensively of turpentine.

Embryonic development is a very complicated physical and chemical performance. Change of detail at any point in its course can have remarkable consequences farther down the line. This is not so surprising, when you recall how heavily bootstrapped the process is. Many of the differences in the way individuals develop are due to differences in environment—oxygen starvation or exposure to thalidomide, for instance. Many other differences are due to differences in genes—not just genes considered in isolation but genes in interaction with other genes, and in interaction with environmental differences. Such a complicated, kaleidoscopic, intricately and reciprocally bootstrapped process as embryonic development is both robust and sensitive. It is robust in that it fights off many potential changes, to produce a living baby against odds that sometimes seem almost overwhelming; at the same time it is sensitive to changes in that no two individuals, not even identical twins, are literally identical in all their features.

And now for the point that this has all been leading up to. To the extent that differences between individuals are due to genes (which may be a large extent or a small one), natural selection can favor some quirks of embryological origami or embryological chemistry and disfavor others. To the extent that your throwing arm is influenced by genes, natural selection can favor it or disfavor it. If being able to throw well has an effect, however slight, on an individual's likelihood of surviving long enough to have children, to the extent that throwing ability is influenced by genes, those genes will have a correspondingly greater chance of winning through to the next generation. Any one individual may die for reasons having nothing to do with his ability to throw. But a gene that tends to make individuals better at throwing when it is present than when it is absent will inhabit lots of bodies, both good and bad, over many generations. From the point of view of the particular gene, the other causes of death will average out. From the gene's perspective, there is only the long-term outlook of the river of DNA flowing down through the generations, only temporarily housed in particular bodies, only temporarily sharing a body with companion genes that may be successful or unsuccessful.

In the long term, the river becomes full of genes that are good at surviving for their several reasons: slightly improving the ability to throw a spear, slightly improving the ability to taste poison, or whatever it may be. Genes that, on average, are less good at surviving—because they tend to cause astigmatic vision in their successive bodies, who are therefore less successful spear throwers; or because they make their successive bodies less attractive and therefore less likely to mate—such genes will tend to disappear from the river of genes. In all this, remember the point we made earlier: the genes that survive in the river will be the ones that are good at surviving in the average environment of the species, and perhaps the most important aspect of this average environment is the other genes of the species; the other genes with which a gene is likely to have to share a body; the other genes that swim through geological time in the same river.

CHAPTER 2

ALL AFRICA AND HER PROGENIES

It is often thought clever to say that science is no more than our modern origin myth. The Jews had their Adam and Eve, the Sumerians their Marduk and Gilgamesh, the Greeks Zeus and the Olympians, the Norsemen their Valhalla. What is evolution, some smart people say, but our modern equivalent of gods and epic heroes, neither better nor worse, neither truer nor falser? There is a fashionable salon philosophy called cultural relativism which holds, in its extreme form, that science has no more claim to truth than tribal myth: science is just the mythology favored by our modern Western tribe. I once was provoked by an anthropologist colleague into putting the point starkly, as follows: Suppose there is a tribe, I said, who believe that the moon is an old calabash tossed into the sky, hanging only just out of reach above the treetops. Do you really claim that our scientific truth—that the moon is about a quarter of a million miles away and a quarter the diameter of the Earth—is no more true than the tribe's calabash? "Yes," the anthropologist said. "We are just brought up in a culture that sees the world in a scientific way. They are brought up to see the world in another way. Neither way is more true than the other."

Show me a cultural relativist at thirty thousand feet and

I'll show you a hypocrite. Airplanes built according to scientific principles work. They stay aloft, and they get you to a chosen destination. Airplanes built to tribal or mythological specifications, such as the dummy planes of the cargo cults in jungle clearings or the beeswaxed wings of Icarus, don't.* If you are flying to an international congress of anthropologists or literary critics, the reason you will probably get there—the reason you don't plummet into a ploughed field—is that a lot of Western scientifically trained engineers have got their sums right. Western science, acting on good evidence that the moon orbits the Earth a quarter of a million miles away, using Western-designed computers and rockets, has succeeded in placing people on its surface. Tribal science, believing that the moon is just above the treetops, will never touch it outside of dreams.

I seldom give a public lecture without a member of the audience brightly coming up with something along the same lines as my anthropologist colleague, and it usually elicits a murmuration of approving nods. No doubt the nodders feel good and liberal and unracist. An even more reliable nod-provoker is "Fundamentally, your belief in evolution comes down to faith, and therefore it's no better than somebody else's belief in the Garden of Eden."

* This is not the first time I have used this knock-down argument, and I must stress that it is aimed strictly at people who think like my colleague of the calabash. There are others who, confusingly, also call themselves cultural relativists although their views are completely different and perfectly sensible. To them, cultural relativism just means that you cannot understand a culture if you try to interpret its beliefs in terms of your own culture. You have to see each of the culture's beliefs in the context of the culture's other beliefs. I suspect that this sensible form of cultural relativism is the original one, and that the one I have criticized is an extremist, though alarmingly common, perversion of it. Sensible relativists should work harder at distancing themselves from the fatuos kind.

Every tribe has had its origin myth—its story to account for the universe, life and humanity. There is a sense in which science does indeed provide the equivalent of this, at least for the educated section of our modern society. Science may even be described as a religion, and I have, not entirely facetiously, published a brief case for science as an appropriate subject for religious-education classes.* (In Britain, religious education is a compulsory part of the school curriculum, unlike in the United States, where it is banned for fear of offending any of the plethora of mutually incompatible faiths.) Science shares with religion the claim that it answers deep questions about origins, the nature of life, and the cosmos. But there the resemblance ends. Scientific beliefs are supported by evidence, and they get results. Myths and faiths are not and do not.

Of all origin myths, the Jewish story of the Garden of Eden is so pervasive in our culture that it has bequeathed its name to an important scientific theory about our ancestry, the theory of "African Eve." I am devoting this chapter to African Eve partly because it will enable me to develop the analogy of the river of DNA but also because I want to contrast her, as a scientific hypothesis, with the legendary matriarch of the Garden of Eden. If I succeed, you will find the truth more interesting, maybe even more poetically moving, than the myth. We begin with an exercise in pure reasoning. Its relevance will become clear soon.

You have two parents, four grandparents, eight great-grandparents and so on. With every generation, the number of ancestors doubles. Go back *g* generations and the number of

**The Spectator* (London), August 6, 1994.

ancestors is 2 multiplied by itself *g* times: 2 to the power *g*. Except that, without leaving our armchair, we can quickly see that it cannot be so. To convince ourselves of this, we have only to go back a little way—say, to the time of Jesus, almost exactly two thousand years ago. If we assume, conservatively, four generations per century—that is, that people breed on average at the age of twenty-five—two thousand years amounts to a mere eighty generations. The real figure is probably more than this (until recent times, many women bred extremely young), but this is only an armchair calculation, and the point is made regardless of such details. Two multiplied by itself 80 times is a formidable number, a 1 followed by 24 noughts, a trillion American trillions. You had a million million million million ancestors who were contemporaries of Jesus, and so did I! But the total population of the world at that time was a fraction of a negligible fraction of the number of ancestors we have just calculated.

Obviously we have gone wrong somewhere, but where? We did the calculation right. The only thing we got wrong was our assumption about doubling up in every generation. In effect, we forgot that cousins marry. I assumed that we each have eight great-grandparents. But any child of a first-cousin marriage has only six great-grandparents, because the cousins' shared grandparents are in two separate ways great-grandparents to the children. "So what?" you may ask. People occasionally marry their cousins (Charles Darwin's wife, Emma Wedgwood, was his first cousin), but surely it doesn't happen often enough to make a difference? Yes it does, because "cousin" for our purposes includes second cousins, fifth cousins, sixteenth cousins and so forth. When you count cousins as distant as that, every marriage is a marriage between cousins. You sometimes hear people boasting of being a distant cousin of the Queen, but it is rather pompous of them, because we are *all* distant cousins of the Queen, and of everybody else, in more ways than can ever be traced. The only thing special about royalty and aristocrats is that they can do the tracing explicitly. As the fourteenth Earl of Home said when taunted about his title by his political opponent, "I suppose Mr. Wilson, when you come to think of it, is the fourteenth Mr. Wilson."

The upshot of all this is that we are much closer cousins of one another than we normally realize, and we have many fewer ancestors than simple calculations suggest. Seeking to get her reasoning along these lines, I once asked a student to make an educated guess as to how long ago her most recent common ancestor with me might have lived. Looking hard at my face, she unhesitatingly replied, in a slow, rural accent, "Back to the apes." An excusable intuitive leap, but it is approximately 10,000 percent wrong. It would suggest a separation measured in millions of years. The truth is that the most recent ancestor she and I shared would probably have lived no more than a couple of centuries ago, probably well after William the Conqueror. Moreover, we were certainly cousins in many different ways simultaneously.

The model of ancestry that led to our erroneously inflated calculation of ancestral numbers was an ever-branching tree, branching and branching again. Turned the other way up, and equally wrong, is a tree model of descent. A typical individual has two children, four grandchildren, eight great-grandchildren and so on down to impossible trillions of descendants a few centuries hence. A far more realistic model of ancestry and descent is the flowing river of genes, which we introduced in the previous chapter. Within its banks, the genes are an ever-rolling stream through time. Currents swirl apart and join again as the genes crisscross down the river of time. Draw out a bucketful at intervals from points spaced out down the length of the river. Pairs of molecules in a bucket will have been companions before, at intervals during their progress down the river, and they will be companions once more. They have also been widely separated in the past, and they will be widely separated again. It is hard to trace the points of contact, but we can be mathematically certain that the contacts happen—mathematically certain that if two genes are out of contact at a particular point, we won't have to travel far in either direction along the river until they touch again.

You may not know that you are a cousin of your husband, but it is statistically likely that you won't have to go far back in your ancestry until you meet a junction with his lineage. Looking in the other direction, toward the future, it might seem obvious that you have a good chance of sharing descendants with your husband or wife. But here is a much more arresting thought. Next time you are with a large group of people—say, in a concert hall or at a football match—look around at the audience and reflect upon the following: if you have any descendants at all in the distant future, there are probably people at the same concert whose hands you could shake as coancestors of your future descendants. Cograndparents of the same children usually know they are coancestors, and this must give them a certain feeling of affinity whether or not they get on personally. They can look at each other and say,"Well, I may not like him much, but his DNA is mingled with mine in our shared grandchild, and we can hope to share descendants into the future, long after we're gone. Surely this creates a bond between us." But my point is that, if you are blessed with distant descendants at all, some of the perfect strangers at the concert hall will probably be your coancestors. You can survey the auditorium and speculate about which individuals, male or female, are destined to share your descendants and which are not. You and I, whoever you are and whatever your color and sex, may well be coancestors. Your DNA may be destined to mingle with mine. Salutations!

Now suppose we travel back in a time machine, perhaps to a crowd in the Colosseum, or farther back, to market day in Ur, or even farther still. Survey the crowd, just as we imagined for our modern concert audience. Realize that you can divide these long-dead individuals into two and only two categories: those who are your ancestors and those who are not. That is obvious enough, but now we come to a remarkable truth. If your time machine has taken you sufficiently far back, you can divide the individuals you meet into those who are ancestors of every human alive in 1995 and those who are the ancestors of nobody alive in 1995. There are no intermediates. Every individual you set eyes on when you step outside your time machine is either a universal human ancestor or not an ancestor of anybody at all.

This is an arresting thought, but it is trivially easy to prove. All you have to do is move your mental time machine back to a ludicrously long time ago: say, to three hundred fifty million years ago, when our ancestors were lobe-finned fishes with lungs, emerging from the water and becoming amphibians. If a particular fish is my ancestor, it is inconceivable that he is not your ancestor too. If he were not, this would imply that the lineage leading to you and the lineage leading to me had independently, without cross-reference, evolved from fish through amphibian, reptile, mammal, primate, ape and hominid, ending up so similar that we can talk to each other and, if we are of opposite sex, mate with each other. What is true of you and me is true of any pair of humans.

We have proved that if we travel sufficiently far back in time, every individual we encounter must be the ancestor either of all of us or of none of us. But how far is sufficiently far? We clearly don't need to go back to lobe-finned fishes—that was the *reductio ad absurdum*—but how far do we have to go back until we come to a universal ancestor of every human alive in 1995? That's a much more difficult question, and it is the one to which I next want to turn. This one cannot be answered from the armchair. We need real information, measurements from the hard world of particular facts.

Sir Ronald Fisher, the formidable English geneticist and mathematician, who could be regarded as Darwin's greatest twentieth-century successor as well as the father of modern statistics, had this to say in 1930:

> It is only the geographical and other barriers to sexual intercourse between different races . . . which prevent the whole of mankind from having had, apart from the last thousand years, a practically identical ancestry. The ancestry of members of the same nation can differ little beyond the last 500 years; at 2000 years the only differences that would seem to remain would be

> those between distinct ethnographic races; these . . . may indeed be extremely ancient; but this could only be the case if for long ages the diffusion of blood between the separated groups was almost non-existent.

In the terms of our river analogy, Fisher is, in effect, making use of the fact that the genes of all the members of one geographically united race are flowing down the same river. But when it came to his actual figures—five hundred years, two thousand years, the antiquity of the separation of different races—Fisher had to have been making educated guesses. The relevant facts were not available in his time. Now, with the molecular-biology revolution, there is an embarrassment of riches. It is molecular biology that has given us the charismatic African Eve.

The digital river is not the only metaphor that has been used. It is tempting to liken the DNA in each one of us to a family Bible. DNA is a very long piece of text, written, as we saw in the previous chapter, in a four-letter alphabet. The letters have been scrupulously copied from our ancestors, and only from our ancestors, with remarkable fidelity even in the case of very remote ancestors. It should be possible, by comparing the texts preserved in different people, to reconstruct their cousinship and work back to a common ancestor. Distant cousins, whose DNA has had more time to diverge—say, Norwegians and Australian aborigines—should differ in a larger number of words. Scholars do this kind of thing with different versions of biblical documents. Unfortunately, in the case of DNA archives, there is a snag. Sex.

Sex is an archivist's nightmare. Instead of leaving ancestral texts intact but for an occasional inevitable error, sex wantonly and energetically wades in and destroys the evidence. No bull ever abused a china shop as sex abuses the DNA archives. There is nothing like it in biblical scholarship. Admittedly, a scholar seeking to trace the origins of, say, the Song of Solomon is aware that it is not quite what it seems. The Song has oddly disjointed passages, suggesting that it is really fragments of several different poems, only some of them erotic, stitched together. It contains errors—mutations—especially in translation. "Take us the foxes, the little foxes, that spoil the vines" is a mistranslation, even though a lifetime's repetition has given it a haunting appeal of its own, which is unlikely to be matched by the more correct "Catch for us the fruit bats, the little fruit bats . . . ":

> For lo, the winter is past, the rain is over and gone. The flowers appear on the earth; the time of the singing of birds is come, and the voice of the turtle is heard in our land.

The poetry is so ravishing that I am reluctant to spoil it by noting that here, too, is an undoubted mutation. Reinsert "dove" after "turtle," as the modern translations correctly but leadenly do, and hear the cadence collapse. But these are minor errors, the inevitable, slight degradations we have to expect when documents are not printed in thousands or etched on high-fidelity computer disks but copied and recopied by mortal scribes from scarce and vulnerable papyri.

But now let sex enter the picture. (No, in the sense I mean, sex does not enter the Song of Songs.) Sex, in the sense I mean, amounts to ripping out half of one document, in the form of randomly chosen fragments, and mixing it with the complementarily butchered half of another document. Unbelievable—vandalistic, even—as it sounds, this is exactly what happens whenever a sex cell is made. For instance, when a man makes a sperm cell, the chromosomes that he inherited from his father pair off with the chromosomes that he inherited from his mother, and great chunks of them change places. A child's chromosomes are an irretrievably scrambled mishmash of its grandparents' chromosomes and so on back to distant ancestors. Of the would-be ancient texts, the letters, perhaps the words, may survive intact down the generations. But chapters, pages, even paragraphs are torn up and recombined with such ruthless efficiency that as a means of tracing history they are almost useless. Where ancestral history is concerned, sex is the great cover-up.

We can use DNA archives to reconstruct history wherever sex is safely out of the picture. I can think of two important examples. One is African Eve, and I'll come to her. The other case is the reconstruction of more remote ancestry—looking at relationships among species rather than within species. As we saw in the previous chapter, sexual mixing takes place only within species. When a parental species buds off a daughter species, the river of genes splits into two branches. After they have diverged for a sufficient time, sexual mixing within each river, far from being a hindrance to the genetic archivist, actually helps in the reconstruction of ancestry and cousinships among species. It is only where within-species cousinships are concerned that sex messes up the evidence. Where between-species cousinships are concerned, sex helps because it tends automatically to ensure that each individual is a good genetic sample of the entire species. It doesn't matter which bucketful you haul out of a well-churned river; it will be representative of the water of that river.

DNA texts taken from representatives of different species have indeed been compared, with great success, letter by letter, to construct family trees of species. It is even possible, according to one influential school of thought, to put dates on the branchings. This opportunity follows from the albeit controversial notion of a "molecular clock": the assumption that mutations in any one region of the genetic text occur at a constant rate per million years. We'll return to the molecular-clock hypothesis in a moment.

The "paragraph" in our genes describing the protein called cytochrome *c* is 339 letters long. Twelve letter changes separate human cytochrome *c* from the cytochrome *c* of horses, our rather distant cousins. Only one cytochrome *c* letter change separates humans from monkeys (our fairly close cousins), one letter change separates horses from donkeys (their very close cousins) and three letter changes separate horses from pigs (their somewhat more distant cousins). Forty-five letter changes separate humans from yeast and the same number separates pigs from yeast. It is not surprising that these numbers should be the same, because as we follow back the river leading to humans, it joins with the river leading to pigs much more recently than their common river joins the river leading to yeast. There is a little slop in these numbers, however. The number of letter changes in cytochrome *c* separating horses from yeast is not forty-five but forty-six. This does not mean that pigs are closer cousins of yeast than horses are. They are exactly equally close to yeast, as are all vertebrates—and, indeed, all animals. Perhaps an extra change crept into the lineage leading to horses since the time of the rather recent ancestor they share with pigs. That is not important. On the whole, the number of cytochrome *c* letter changes separating pairs of creatures is pretty much what we'd expect from previous ideas of the branching pattern of the evolutionary tree.

The molecular-clock theory, as noted, holds that the rate of change of a given piece of text per million years is roughly fixed. Of the forty-six cytochrome *c* letter changes separating horses from yeast, it is assumed that about half of them occurred during evolution from the common ancestor to modern horses and about half of them occurred during evolution from the common ancestor to modern yeast (obviously, the two evolutionary pathways have taken exactly the same number of millions of years to accomplish). At first this seems a surprising thing to assume. After all, it is pretty likely that the common ancestor resembled yeast more than it resembled a horse. The reconciliation lies in the assumption, increasingly accepted since it was first championed by the eminent Japanese geneticist Motoo Kimura, that the greater part of genetic texts can change freely without the text's meaning being affected.

A good analogy is varying the typeface in a printed sentence. "A **horse** is *a* **mammal**." "A **yeast is** *a* **fungus.**" The meaning of these sentences comes through loud and clear, even though every word is printed in a different font. The molecular clock ticks away in the equivalent of meaningless font changes, as the millions of years go by. The changes that are subject to natural selection and that describe the difference between a horse and a yeast—the changes in *meaning* of the sentences—are the tip of the iceberg.

Some molecules have a higher clock rate than others.

Cytochrome *c* evolves relatively slowly: about one letter change every twenty-five million years. This is probably because cytochrome *c*'s vital importance to an organism's survival depends critically upon its detailed shape. Most changes in such a shape-critical molecule are not tolerated by natural selection. Other proteins, such as those called fibrinopeptides, although they are important, work equally well in lots of variant forms. The fibrinopeptides are used in blood clotting, and you can change most of their details without harming their clottability. The mutation rate in these proteins is about one change every six hundred thousand years, a rate more than forty times faster than that for cytochrome *c*. Fibrinopeptides, therefore, are no good for reconstructing ancient ancestry, although they are useful for reconstructing more recent ancestry—for example, within the mammals. There are hundreds of different proteins, each changing at its own characteristic rate per million years and each independently usable for reconstructing family trees. They all yield pretty much the same family tree—which, by the way, is rather good evidence, if evidence were needed, that the theory of evolution is true.

We came into this discussion from the realization that sexual mixing messes up the historical record. We distinguished two ways in which the effects of sex could be escaped. We've just dealt with one of them, following from the fact that sex does not mix genes between species. This opens up the possibility of using DNA sequences to reconstruct remotely ancient family trees of our ancestors that lived long before we became recognizably human. But we've already agreed that if we go back that far we humans are all definitely descended from the same single individual anyway. We wanted to find out how recently we could still claim common descent with all other humans. To discover that, we have to turn to a different kind of DNA evidence. This is where African Eve comes into the story.

African Eve is sometimes called Mitochondrial Eve. Mitochondria are tiny, lozenge-shaped bodies swarming by the thousands in each one of our cells. They are basically hollow but with a complicated interior structure of membranous baffles. The area afforded by these membranes is much larger than you'd think from the outside appearance of mitochondria, and it is used. The membranes are the production lines of a chemical factory—more precisely, a power station. A carefully controlled chain reaction is strung out along the membranes—a chain reaction involving more stages than those in any human chemical factory. The result is that energy, originating in food molecules, is released in controlled steps and stored in reusable form for burning later, wherever it is needed, anywhere in the body. Without our mitochondria, we'd die in a second.

That's what mitochondria do, but we are here more concerned with where they come from. Originally, in ancient evolutionary history, they were bacteria. This is the remarkable theory championed, by the redoubtable Lynn Margulis of the University of Massachusetts at Amherst, from heterodox origins through grudging interest to triumphant near-universal acceptance today. Two billion years ago, the remote ancestors of mitochondria were free-living bacteria. Together with other bacteria of different kinds, they took up residence inside larger cells. The resulting community of ("prokaryotic") bacteria became the large ("eukaryotic") cell we call our own. Each one of us is a community of a hundred million million mutually dependent eukaryotic cells. Each one of those cells is a community of thousands of specially-tamed bacteria, entirely enclosed within the cell, where they multiply as bacteria will. It has been calculated that if all the mitochondria in a single human body were laid end to end, they would girdle the Earth not once but two thousand times. A single animal or plant is a vast community of communities packed in interacting layers, like a rain forest. As for a rain forest itself, it is a community seething with perhaps ten million species of organisms, every individual member of every species being itself a community of communities of domesticated bacteria. Not only is Dr. Margulis's theory of origins—the cell as an enclosed garden of bacteria—incomparably more inspiring, exciting and uplifting than the story of the Garden of Eden. It has the additional advantage of being almost certainly true.

Like most biologists, I now assume the truth of the Margulis theory, and in this chapter I mention it only to follow up a particular implication: mitochondria have their own DNA, which is confined to a single ring chromosome, as in other bacteria. And now for the point that this has all been leading up to. Mitochondrial DNA does not participate in any sexual mixing, either with the main "nuclear" DNA of the body or with the DNA of other mitochondria. Mitochondria, like many bacteria, reproduce simply by dividing. Whenever a mitochondrion divides into two daughter mitochondria, each daughter gets an identical copy—give or take the odd mutation—of the original chromosome. Now you see the beauty of this, from our point of view as long-distance genealogists. We found that where our ordinary DNA texts are concerned, in every generation sex scrambles the evidence, confusing the contributions from paternal and maternal lines. Mitochondrial DNA is blessedly celibate.

We get our mitochondria from our mother only. Sperms are too small to contain more than a few mitochondria; they have just enough to provide the energy to power their tails as they swim toward the egg, and these mitochondria are cast away with the tail when the sperm head is absorbed in the egg at fertilization. The egg is massive by comparison, and its huge, fluid-filled interior contains a rich culture of mitochondria. This culture seeds the child's body. So whether you are female or male, your mitochondria are all descended from an initial inoculum of your mother's mitochondria. Whether you are male or female, your mitochondria are all descended from your maternal grandmother's mitochondria. None from your father, none from either grandfather, none from your paternal grandmother. The mitochondria constitute an independent record of the past, uncontaminated by the main nuclear DNA, which is equally likely to come from each of four grandparents, each of eight great-grandparents and so on back.

Mitochondrial DNA is uncontaminated, but it is not immune to mutation—to random errors in copying. Indeed, it mutates at a higher rate than our "own" DNA, because (as is the case with all bacteria) it lacks the sophisticated proofreading machinery our cells have evolved over the eons. There will be a few differences between your mitochondrial DNA and mine, and the number of differences will be a measure of how far back our ancestors diverged. Not *any* of our ancestors, but our ancestors in the female female female . . . line. If your mother happens to be a purebred native Australian, or a purebred Chinese, or a purebred !Kung San of the Kalahari, there will be rather a lot of differences between your mitochondrial DNA and mine. It doesn't matter who your father is: he can be an English marquess or a Sioux chieftain, for all the difference it makes to your mitochondria. And the same goes for any of your male ancestors, ever.

So there is a separate mitochondrial Apocrypha, handed down alongside the main family Bible but with the great virtue of going down the female line only. This is not a sexist point; it would be just as good if it came down through the male line only. The virtue lies in its intactness, in its not being chopped and merged in every generation. Consistent descent via either sex but not both is what we, as DNA genealogists, need. The Y chromosome which, like a surname, is handed down the male line only, would in theory be just as good, but it contains too little information to be useful. The mitochondrial Apocrypha is ideal for dating common ancestors within one species.

Mitochondrial DNA has been exploited by a group of researchers associated with the late Allan Wilson in Berkeley, California. In the 1980s, Wilson and his colleagues sampled the sequences from 135 living women drawn from all around the world—Australian aborigines, New Guinea highlanders, Native Americans, Europeans, Chinese and representatives of various peoples in Africa. They looked at the numbers of letter differences separating each woman from each other woman. They gave these numbers to a computer and asked it to construct the most parsimonious family tree it could find. "Parsimonious" here means doing away as much as possible with the need to postulate coincidence. This requires some explaining.

Think back to our earlier discussion of horses, pigs and

yeast, and the analysis of cytochrome *c* letter sequences. You remember that horses differ from pigs in only three such letters, pigs differ from yeast in forty-five letters, and horses differ from yeast in forty-six letters. We made the point that, theoretically, since horses and pigs are connected to each other by a relatively recent common ancestor, they should be exactly the same distance from yeast. The difference between forty-five and forty-six is an anomaly, something that in an ideal world would not be there. It may be due to an additional mutation on the route to horses or a reverse mutation on the route to pigs.

Now, absurd as such an idea is in reality, it is theoretically conceivable that pigs are really closer to yeast than they are to horses. It is theoretically possible that pigs and horses have evolved their close resemblance to one another (their cytochrome *c* texts are only three letters apart, and their bodies are basically built to an almost identical mammalian pattern) by massive coincidence. The reason we don't believe this is that the ways in which pigs resemble horses vastly outnumber the ways in which pigs resemble yeast. Admittedly, there is a single DNA letter in which pigs appear closer to yeast than to horses, but this is swamped by millions of resemblances going the other way. The argument is one of parsimony. If we assume that pigs are close to horses, we need to accommodate only one coincidental resemblance. If we try to assume that pigs are close to yeast, we have to postulate a prodigiously unrealistic concatenation of independently acquired coincidental resemblances.

In the cases of horses, pigs and yeast, the parsimony argument is too overwhelming to be in doubt. But in the mitochondrial DNA of different human races there is nothing overwhelming about the resemblances. Parsimony arguments still apply, but they are slight, quantitative arguments, not massive, knock-down arguments. Here's what the computer, in theory, has to do. It has to make a list of all possible family trees relating the 135 women. It then examines this set of possible trees and picks out the most parsimonious one—that is, the one that minimizes the number of coincidental resemblances. We must accept that even the best tree will probably force us to accept a few little coincidences, just as we were forced to accept the fact that, with regard to one DNA letter, yeasts are closer to pigs than to horses. But—in theory, at least—the computer should be able to take that in its stride and announce to us which of the many possible trees is the most parsimonious, the least coincidence-ridden.

That is in theory. In practice, there is a snag. The number of possible trees is greater than you, or I, or any mathematician, can possibly imagine. For horse, pig and yeast there are only three possible trees. The obviously correct one is [[pig horse] yeast], with pig and horse nested together inside the innermost brackets and yeast as the unrelated "outgroup." The other two theoretical trees are *[[pig yeast] horse]* and *[[horse yeast] pig]*. If we add a fourth creature—say, squid—the number of trees goes up to twelve. I won't list all twelve, but the true (most parsimonious) one is *[[[pig horse] squid] yeast]*. Again, pig and horse, as close relatives, are cosily nested together in the innermost brackets. Squid is the next to join the club, having a more recent ancestor with the pig/horse lineage than yeast does. Any of the eleven other trees—for instance, *[[pig squid] [horse yeast]]*—is definitely less parsimonious. It is highly improbable that pig and horse could have independently evolved their numerous resemblances if pig were really a closer cousin to squid and horse were really a closer cousin to yeast.

If three creatures yield three possible trees, and four creatures yield twelve possible trees, how many possible trees could be constructed for a hundred and thirty-five women? The answer is such a risibly large number that there is no point in writing it out. If the largest and fastest computer in the world were set to work listing all the possible trees, the end of the world would be upon us before the computer had made a perceptible dent in the task.

Nevertheless, the problem is not hopeless. We are used to taming impossibly large numbers by judicious sampling techniques. We can't count the number of insects in the Amazon Basin, but we can estimate the number by sampling small plots dotted at random through the forest and assuming that these plots are representative. Our computer can't examine all possible trees uniting the 135 women, but it can pull out random samples from the set of all possible trees. If, whenever you draw a sample from the gigabillions of possible trees, you notice that the most parsimonious members of the sample have certain features in common, you can conclude that probably the most parsimonious of all the trees has the same features.

This is what people have done. But it isn't necessarily obvious what is the best way to do it. Just as entomologists might disagree over the most representative way to sample the Brazilian rain forest, so DNA genealogists have used different sampling methods. And unfortunately the results don't always agree. Nevertheless, for what they are worth, I'll present the conclusions the Berkeley group reached in their original analysis of human mitochondrial DNA. Their conclusions were extremely interesting and provocative. According to them, the most parsimonious tree turns out to be firmly rooted in Africa. What this means is that some Africans are more distantly related to other Africans than to anybody in the whole of the rest of the world. The whole of the rest of the world—Europeans, Native Americans, Australian aboriginals, Chinese, New Guineans, Inuits, and all—form one relatively close group of cousins. Some Africans belong in this close group. But other Africans don't. According to this analysis, the most parsimonious tree looks like this: [some Africans [other Africans [yet other Africans [yet other Africans and everybody else]]]]. They therefore concluded that the grand ancestress of all of us lived in Africa: "African Eve." As I have said, this conclusion is controversial. Others have claimed that equally parsimonious trees can be found in which the outermost branches occur outside Africa. They also claim that the Berkeley group obtained the particular results they did partly because of the order in which their computer looked at the possible trees. Obviously, order of looking ought not to matter. Probably most experts would still put their money on Mitochondrial Eve's being African, but they wouldn't do so with any great confidence.

The second conclusion of the Berkeley group is less controversial. No matter where Mitochondrial Eve lived, they were able to estimate when. It is known how fast mitochondrial DNA evolves; you can therefore put an approximate date on each of the branch points on the tree of divergence of mitochondrial DNA. And the branch point that unites all womankind—the birth date of Mitochondrial Eve—is between a hundred fifty thousand and a quarter of a million years ago.

Whether Mitochondrial Eve was an African or not, it is important to avoid a possible confusion with another sense in which it is undoubtedly true that our ancestors came out of Africa. Mitochondrial Eve is a recent ancestor of all modern humans. She was a member of the species *Homo sapiens*. Fossils of much earlier hominids, *Homo erectus,* have been found outside as well as inside Africa. The fossils of ancestors even more remote than *Homo erectus*, such as *Homo habilis* and various species of *Australopithecus* (including a newly discovered one more than four million years old), have been found only in Africa. So if we are the descendants of an African diaspora within the last quarter of a million years, it is the second African diaspora. There was an earlier exodus, perhaps a million and a half years ago, when *Homo erectus* meandered out of Africa to colonize parts of the Middle East and Asia. The African Eve theory is claiming not that these earlier Asians didn't exist but that they leave no surviving descendants. Whichever way you look at it, we are all, if you go back two million years, Africans. The African Eve theory is claiming in addition that we surviving humans are all Africans if you go back only a few hundred thousand years. It would be possible, if new evidence supported it, to trace all modern mitochondrial DNA back to an ancestress outside Africa ("Asian Eve," say) while at the same time agreeing that our more remote ancestors are to be found only in Africa.

Let's assume, for the moment, that the Berkeley group is right, and examine what their conclusion does and does not mean. The "Eve" sobriquet has had unfortunate consequences. Some enthusiasts have run away with the idea that she must have been a lonely woman, the only woman on Earth, the ultimate genetic bottleneck, even a vindication of Genesis! This is a complete misunderstanding. The correct claim is not that she was the only woman on Earth, nor even that the population was relatively small during her time. Her companions, of both sexes, may have been both numerous and fecund. They may still have numerous descendants alive today. But all descendants of their mitochondria have died out, because their link with us passes, at some point, through a male. In the same way, a noble surname (surnames are linked to Y chromosomes and pass down the male-only line in exact mirror image to mitochondria) can die out, but this doesn't mean that possessors of the surname have no descendants. They may have numerous descendants via pathways other than the male-only pathway. The correct claim is only that Mitochondrial Eve is the most recent woman of whom it can be said that all modern humans are descended from her in the female-only line. There *has* to be *a* woman of whom this claim can be made. The only argument is over whether she lived here rather than there, at this time rather than at that time. The fact that she did live, in some place and at some time, is certain.

Here is a second misunderstanding—a more common one, which I have heard perpetrated even by leading scientists working in the field of mitochondrial DNA. This is the belief that Mitochondrial Eve is our most recent common ancestor. It is based on a confusion between "most recent common ancestor" and "most recent common ancestor in the purely female line." Mitochondrial Eve is our most recent common ancestor in the purely female line, but there are lots of other ways of being descended from people than in the female line. Millions of other ways. Go back to our calculations of numbers of ancestors (forgetting the complication of cousin marriage, which was the point of the argument before). You have eight great-grandparents but only one of them is in the purely female line. You have sixteen great-great-grandparents but only one of *them* is in the purely female line. Even allowing that cousin marriage reduces the number of ancestors in a given generation, it is still true that there are far, far, far more ways of being an ancestor than just in the female-only line. As we follow our genetic river back through remote antiquity, there were probably lots of Eves and lots of Adams—focal individuals, of whom it is possible to say that all 1995's people are descended from her, or him. Mitochondrial Eve is only one of these. There is no particular reason to think that of all these Eves and Adams, Mitochondrial Eve is the most recent. On the contrary. She is defined in a *particular* way: we are descended from her via a particular pathway through the river of descent. The number of possible pathways to set alongside the female-only pathway is so large that it is mathematically highly unlikely that Mitochondrial Eve is the most recent of these many Eves and Adams. It is special among pathways in one way (being female-only). It would be a remarkable coincidence if it were special among pathways in another way (being the most recent).

An additional point of mild interest is that our most recent common ancestor is somewhat more likely to have been an Adam than an Eve. Harems of females are more likely to occur than harems of males, if only because males are physically capable of having hundreds of children, even thousands. *The Guinness Book of Records* puts the record at over a thousand, achieved by Moulay Ishmael the Bloodthirsty. (Incidentally, Moulay Ishmael might well be adopted by feminists as a general symbol of macho unpleasantness. It is said that his method of mounting a horse was to draw his sword and leap into the saddle, achieving quick release by simultaneously decapitating the slave who held the bridle. Implausible as this is, the fact that the legend comes down to us, together with his reputation for having killed ten thousand men with his own hand, perhaps gives an idea of the kinds of qualities that were admired among men of his type.) Females, even under ideal conditions, cannot have more than a couple of tens of children. A female is more likely than a male to have the average number of children. A few males may have a ludicrously greedy share of the children, which means that other males must have none. If anybody fails to reproduce altogether, it is more likely to be a male than a female. And if anybody garners a disproportionate posterity, it is also likely to be a male. This goes for the most recent common ancestor of all humanity, who is therefore more likely to have been an Adam than an Eve. To take an extreme example, who is more likely to be the ancestor of all present-day Moroccans, Moulay Ishmael the Bloodthirsty or any *one* woman in his unfortunate harem?

We may come to the following conclusions: First, it is necessarily certain that there existed one female, whom we may call Mitochondrial Eve, who is the most recent common ancestor of all modern humans down the female-only pathway. It is also certain that there existed one person, of unknown sex, whom we may call the Focal Ancestor, who is the most recent common ancestor of all modern humans down any pathway. Third, although it is possible that Mitochondrial Eve and the Focal Ancestor are one and the same, it is vanishingly unlikely that this is so. Fourth, it is somewhat more likely that the Focal Ancestor was a male than a female. Fifth, Mitochondrial Eve very probably lived less than a quarter of a million years ago. Sixth, there is disagreement over where Mitochondrial Eve lived, but the balance of informed opinion still favors Africa. Only conclusions five and six depend upon inspection of scientific evidence. The first four can all be worked out by armchair reasoning from common knowledge.

But I said that ancestors hold the key to understanding life itself. The story of African Eve is a parochial, human microcosm of a grander and incomparably more ancient epic. We shall again have recourse to the metaphor of the river of genes, our river out of Eden. But we shall follow it back through a time scale incommensurably older than the legendary Eve's thousands of years and African Eve's hundreds of thousands. The river of DNA has been flowing through our ancestors in an unbroken line that spans not less than three thousand million years.

CHAPTER 3

DO GOOD BY STEALTH

Creationism has enduring appeal, and the reason is not far to seek. It is not, at least for most of the people I encounter, because of a commitment to the literal truth of Genesis or some other tribal origin story. Rather it is that people discover for themselves the beauty and complexity of the living world and conclude that it "obviously" must have been designed. Those creationists who recognize that Darwinian evolution provides at least some sort of alternative to their scriptural theory often resort to a slightly more sophisticated objection. They deny the possibility of evolutionary intermediates. "X *must* have been designed by a Creator," people say, "because half an X would not work at all. All the parts of X must have been put together simultaneously; they could not have gradually evolved." For instance, on the day I began writing this chapter I happened to receive a letter. It was from an American minister who had been an atheist but was converted by reading an article in *National Geographic.* Here is an extract from the letter:

> The article was about the amazing adaptations that orchids have made to their environments in order to propagate successfully. As I read I was particularly intrigued by the reproductive strategy of one species, which involved the cooperation of a male

> wasp. Apparently the flower resembled very closely the female of this species of wasp, including having an opening in the proper place, so that the male wasp might just reach, by copulating with the flower, the pollen produced by the blossom. Flying on to the next flower the process would be repeated, and thus cross-pollination take place. And what made the flower attractive to the wasp in the first place was that it emitted pheromones [specific chemical attractants much used by insects to bring the sexes together] identical to the female of that species of wasp. With some interest I studied the accompanying picture for a minute or so. Then, with a terrific sense of shock, I realized that in order for that reproductive strategy to have worked at all, it had to be perfect the first time. No incremental steps could account for it, for if the orchid did not look like and smell like the female wasp, and have an opening suitable for copulation with the pollen within perfect reach of the male wasp's reproductive organ, the strategy would have been a complete failure.
>
> I will never forget the sinking feeling that overwhelmed me, because it became clear to me in that minute that some kind of God in some kind of fashion must exist, and have an ongoing relationship with the processes by which things come into being. That in short, the creator God was not some antediluvian myth, but something real. And, most reluctantly, I also saw at once that I must search to find out more about that God.

Others, no doubt, come to religion by different routes, but certainly many people have had an experience similar to the one that changed the life of this minister (whose identity I shall withhold out of good manners). They have seen, or read about, some marvel of nature. This has, in a general way, filled them with awe and wonderment, spilling over into reverence. More specifically, like my correspondent, they have decided that this particular natural phenomenon—a spider's web, or an eagle's eye or wing, or whatever it is—cannot have evolved by gradual stages, because the intermediate, half-formed stages could not have been good for anything. The purpose of this chapter is to destroy the argument that complicated contrivances have to be perfect if they are to work at all. Incidentally, orchids were among Charles Darwin's favorite examples, and he devoted a whole book to showing how the principle of gradual evolution by natural selection triumphantly meets the ordeal of explaining "The Various Contrivances by which Orchids are Fertilised by Insects."

The key to the minister's argument lies in the assertion that "in order for that reproductive strategy to have worked at all, it had to be perfect the first time. No incremental steps could account for it." The same argument could be made—frequently has been made—for the evolution of the eye, and I'll return to this in the course of the chapter.

What always impresses me whenever I hear this kind of argument is the confidence with which it is asserted. How, I want to ask the minister, can you be so *sure* that the wasp-mimicking orchid (or the eye, or whatever) wouldn't work unless every part of it was perfect and in place? Have you, in fact, given the matter a split-second's thought? Do you actually know the first thing about orchids, or wasps, or the eyes with which wasps look at females and orchids? What emboldens you to assert that wasps are so hard to fool that the orchid's resemblance would have to be perfect in all dimensions in order to work?

Think back to the last time you were fooled by some chance resemblance. Perhaps you raised your hat to a stranger in the street, mistaking her for an acquaintance. Film stars have stand-in stuntmen or stuntwomen to fall off horses or jump off cliffs in their stead. The stuntman's resemblance to the star is usually extremely superficial, but in the fleeting action shot it is enough to fool an audience. Human males are roused to lust by pictures in a magazine. A picture is just printing ink on paper. It is two-dimensional, not three. The image is only a few inches high. It may be a crude caricature consisting of a few lines, rather than a life-like representation. Yet it can still arouse a man to erection. Perhaps a fleeting view of a female is all a fast-flying wasp can expect to get before attempting to copulate with her. Perhaps male wasps notice only a few key stimuli anyway.

There is every reason to think that wasps might be even easier to fool than humans. Sticklebacks certainly are, and fish have bigger brains and better eyes than wasps. Male sticklebacks have red bellies, and they will threaten not only other males but also crude dummies with red "bellies." My old maestro, the Nobel Prize–winning ethologist Niko Tinbergen, told a famous story about a red mail van that drove past the window of his laboratory, and how all the male sticklebacks rushed to the window side of their tanks and vigorously threatened it. Female sticklebacks that are ripe with eggs have conspicuously swollen bellies. Tinbergen found that an extremely crude, vaguely elongated, silvery dummy, looking nothing like a stickleback to our eyes but possessed of a well-rounded "belly," evoked full mating behavior from males. More recent experiments in the school of research founded by Tinbergen have shown that a so-called sex bomb—a pear-shaped object, rounded plumpness personified but not elongated and not fishlike by any stretch of the (human) imagination—was even more effective in arousing the lusts of the male stickleback. The stickleback "sex bomb" is a classic example of a supernormal stimulus—a stimulus even more effective than the real thing. As another example, Tinbergen published a picture of an oystercatcher trying to sit on an egg the size of an ostrich egg. Birds have bigger brains and better eyesight than fish—and *a fortiori* than wasps—yet oystercatchers apparently "think" that an ostrich-sized egg is a superlative object for incubation.

Gulls, geese and other ground-nesting birds have a stereotyped response to an egg that has rolled out of the nest. They reach over and roll it back in with the underside of their bill. Tinbergen and his students showed that gulls will do this not just to their own eggs but to hens' eggs and even wooden cylinders or cocoa tins discarded by campers. Baby herring gulls get their food by begging from their parents; they peck at the red spot on the parent's bill, stimulating the parent to regurgitate some fish from its bulging crop. Tinbergen and a colleague showed that crude cardboard dummies of a parent's head are very effective in provoking begging behavior from the young. All that is really necessary is a red spot. As far as the baby gull is concerned, its parent is a red spot. It may well see the rest of its parent, but that doesn't seem to be important.

This apparently restricted vision is not confined to baby gulls. Adult black-headed gulls are conspicuous because of their dark face masks. Tinbergen's student Robert Mash investigated the importance of this to other adults by painting wooden dummy gull heads. Each head was stuck on the end of a wooden rod attached to electric motors in a box so that, by remote control, Mash could raise or lower the head and turn it left or right. He would bury the box near a gull nest and leave it with the head safely out of sight beneath the sand. Then, day after day, he would visit a blind near the nest and observe the nesting gulls' reaction to the dummy head when it was raised and turned this way or that. The birds responded to the head and to its turning just as though it were a real gull, yet it was only a mock-up on the end of a wooden rod, without any body, without legs or wings or tail, silent and without movement apart from a pretty unlifelike, robotic rising, rotating and lowering. To a black-headed gull, it seems, a threatening neighbor is little more than a disembodied black face. No body, or wings, or anything else seem to be necessary.

Just to get into the blind to observe the birds, Mash, like generations of ornithologists before him and since, exploited a long-known limitation of the bird nervous system: birds are not natural mathematicians. Two of you go to the blind, and only one of you leaves it. Without this trick, the birds would be wary of the blind, "knowing" that somebody had entered it. But if they see one person leave, they "assume" that both have left. If a bird can't tell the difference between one person and two, is it all that surprising that a male wasp might be fooled by an orchid that bore a less than perfect resemblance to a female?

One more bird story along these lines, and it is a tragedy. Turkey mothers are fierce protectors of their young. They need to protect them against nest marauders like weasels or scavenging rats. The rule of thumb a turkey mother uses to recognize nest robbers is a dismayingly brusque one: In the vicinity of your nest, attack anything that moves, *unless* it makes a noise like a baby turkey. This was discovered by an Austrian zoologist named Wolfgang Schleidt. Schleidt once had a mother turkey that savagely killed all her own babies. The reason was woefully simple: she was deaf. Predators, as far as the turkey's nervous system is concerned, are defined as moving objects that don't emit a baby's cry. These baby turkeys, though they looked like baby turkeys, moved like baby turkeys, and ran trustingly to their mother like baby turkeys, fell victim to the mother's restricted definition of a "predator." She was protecting her own children against themselves, and she massacred them all.

In an insect echo of the tragic story of the turkey, certain of the sensory cells in honeybee antennae are sensitive to only one chemical, oleic acid. (They have other cells sensitive to other chemicals.) Oleic acid is given off by decaying bee corpses, and it triggers the bees' "undertaker behavior," the removal of dead bodies from the hive. If an experimenter paints a drop of oleic acid on a live bee, the wretched creature is dragged off, kicking and struggling and obviously very much alive, to be thrown out with the dead.

Insect brains are much smaller than turkey brains or human brains. Insect eyes, even the big compound eyes of dragonflies, possess a fraction of the acuity of our eyes or bird eyes. Quite apart from this, it is known that insect eyes see the world in a completely different way from our eyes. The great Austrian zoologist Karl von Frisch discovered as a young man that they are blind to red light but they can see—and see as its own distinct hue—ultraviolet light, to which we are blind. Insect eyes are much preoccupied with something called "flicker," which seems—at least to a fast-moving insect—to substitute partially for what we would call "shape." Male butterflies have been seen to "court" dead leaves fluttering down from the trees. We see a female butterfly as a pair of large wings flapping up and down. A flying male butterfly sees her, and courts her, as a concentration of "flicker." You can fool him with a stroboscopic lamp, which doesn't move but just flashes on and off. If you get the flickering rate right, he will treat it as if it were another butterfly flapping its wings at that rate. Stripes, to us, are static patterns. To an insect as it flies past, stripes appear as "flicker" and can be mimicked with a stroboscopic lamp flashing at the right rate. The world as seen through an insect's eyes is so alien to us that to make statements based on our own experience when discussing how "perfectly" an orchid needs to mimic a female wasp's body is human presumption.

Wasps themselves were the subject of a classic experiment, originally done by by the great French naturalist Jean-Henri Fabre and repeated by various other workers, including members of Tinbergen's school. The female digger wasp returns to her burrow carrying her stung and paralyzed prey. She leaves it outside the burrow while she enters, apparently to check that all is well before she reappears to drag the prey in. While she is in the burrow, the experimenter moves the prey a few inches away from where she left it. When the wasp resurfaces, she notices the loss and quickly relocates the prey. She then drags it back to the burrow entrance. Only a few seconds have passed since she inspected the inside of the burrow. We think that there is really no good reason why she should not proceed to the next stage in her routine, drag the prey inside and be done with it. But her program has been reset to an earlier stage. She dutifully leaves the prey outside the burrow again and goes inside for yet another inspection. The experimenter may repeat this charade forty times, until he gets bored. The wasp behaves like a washing machine that has been set back to an early stage in its program and doesn't "know" that it has already washed those clothes forty times without a break. The distinguished computer scientist Douglas Hofstadter has adopted a new adjective, "sphexish," to label such inflexible, mindless automatism. (*Sphex* is the name of one representative genus of digger wasp.) At least in some respects, then, wasps are easy to fool. It is a very different kind of fooling from that engineered by the orchid. Nevertheless, we must beware of using human intuition to conclude that "in order for that reproductive strategy to have worked at all, it had to be perfect the first time."

I may have done my work too well in persuading you that wasps are likely to be easy to fool. You may be nurturing a suspicion almost opposite to that of my ordained correspondent. If insect eyesight is so poor, and if wasps are so easy to fool, why does the orchid bother to make its flower as wasp-like as it is? Well, wasp eyesight is not always so poor. There are situations in which wasps seem to see quite well: when they are locating their burrow after a long hunting flight, for instance. Tinbergen investigated this with the bee-hunting digger wasp, *Philanthus.* He would wait until a wasp was down in her burrow. Before she reemerged, Tinbergen would hastily place some "landmarks" around the entrance to the burrow—say, a twig and a pinecone. He would then retreat and wait for the wasp to fly out. After she did so, she flew two or three circles around the burrow, as though taking a mental photograph of the area, then flew off to seek her prey. While she was gone, Tinbergen would move the twig and the pinecone to a location a few feet away. When the wasp returned, she missed her burrow and instead dived into the sand at the appropriate point relative to the new positions of the twig and the pinecone. Again, the wasp has been "fooled," in a sense, but this time she earns our respect for her eyesight. It looks as though "taking a mental photograph" was indeed what she was doing on her preliminary circling flight. She seems to have recognized the pattern, or "gestalt," of the twig and the pinecone. Tinbergen repeated the experiment many times, using different kinds of landmarks, such as rings of pinecones, with consistent results.

Now here's an experiment of Tinbergen's student Gerard Baerends that contrasts impressively with Fabre's "washing machine" experiment. Baerends' species of digger wasp, *Ammophila campestris* (a species also studied by Fabre), is unusual in being a "progressive provisioner." Most digger wasps provision their burrow and lay an egg, then seal up the burrow and leave the young larva to feed on its own. *Ammophila* is different. Like a bird, it returns daily to the burrow to check on the larva's welfare, and gives it food as needed. Not particularly remarkable, so far. But an individual female *Ammophila* will have two or three burrows on the go at any one time. One burrow will have a relatively large, nearly grown larva; one a small, new-laid larva; and one, perhaps, a larva of intermediate age and size. The three naturally have different food requirements, and the mother tends them accordingly. By a painstaking series of experiments involving the swapping of nest contents, Baerends was able to show that mother wasps do indeed take account of the different food requirements of each nest. This seems clever, but Baerends found that it is also not clever, in a very odd, alien way. The mother wasp, first thing each morning, makes a round of inspection of all her active burrows. It is the state of each nest at the time of the dawn inspection that the mother measures and that influences her provisioning behavior for the rest of the day. Baerends could swap nest contents as often as he pleased after the dawn inspection, and it made no difference to the mother wasp's provisioning behavior. It was as though she switched on her nest-assessing apparatus only for the duration of the dawn inspection round and then switched it off, to save electricity for the rest of the day.

On the one hand, this story suggests that there is sophisticated equipment for counting, measuring, and even calculating, in the mother wasp's head. It now becomes easy to believe that the wasp brain would indeed be fooled only by a thoroughly detailed resemblance between orchid and female. But at the same time, Baerends' story suggests a capacity for selective blindness and a foolability that are all of a piece with the washing-machine experiment, and make it believable that a crude resemblance between orchid and female might well be sufficient. The general lesson we should learn is never to use human judgment in assessing such matters. Never say, and never take seriously anybody who says, "I cannot believe that so-and-so could have evolved by gradual selection." I have dubbed this kind of fallacy "the Argument from Personal Incredulity." Time and again, it has proved the prelude to an intellectual banana-skin experience.

The argument I am attacking is the one that says: gradual evolution of so-and-so couldn't have happened, because so-and-so "obviously" has to be perfect and complete if it is to work at all. So far, in my reply, I've made much of the fact that wasps and other animals have a very different view of the world from our own, and in any case even we are not difficult to fool. But there are other arguments I want to develop that are even more convincing and more general. Let's use the word "brittle" for a device that must be perfect if it is to work at all—as my correspondent alleged of wasp-mimicking orchids. I find it significant that it is actually quite hard to think of an unequivocally brittle device. An airplane is not brittle, because although we'd all prefer to entrust our lives to a Boeing 747 complete with all its myriad parts in perfect working order, a plane that has lost even major pieces of equipment, like one or two of its engines, can still fly. A microscope is not brittle, because although an inferior one gives a fuzzy and ill-lit image, you can still see small objects better with it than you could with no microscope at all. A radio is not brittle; if it is deficient in some respect, it may lose fidelity and its sounds may be tinny and distorted, but you can still make out what the words mean. I have been staring out of the window for ten minutes trying to think of a single really good example of a brittle man-made device, and I can come up with only one: the arch. An arch has a certain near-brittleness in the sense that, once its two sides have come together, it has great strength and stability. But before the two sides come together, neither side will stand up at all. An arch has to be built with the aid of some sort of scaffolding. The scaffolding provides temporary support until the arch is complete; then it can be removed and the arch remains stable for a very long time.

There is no reason in human technology why a device should not in principle be brittle. Engineers are at liberty to design, on their drawing boards, devices that, if half-complete, would not work at all. Even in the field of engineering, however, we are hard put to find a genuinely brittle device. I believe that this is even more true of living devices. Let's look at some of the allegedly brittle devices from the living world that creationist propaganda has served up. The example of the wasp and the orchid is only one example of the fascinating phenomenon of mimicry. Large numbers of animals and some plants gain an advantage because of their resemblance to other objects, often other animals or plants. Almost every aspect of life has somewhere been enhanced or subverted by mimicry: catching food (tigers and leopards are nearly invisible as they stalk their prey in sun-dappled woodland; angler fish resemble the sea bottom on which they sit, and they lure their prey with a long "fishing rod," on the end of which is a bait that mimics a worm; *femmes fatales* fireflies mimic the courtship flash patterns of another species, thereby luring males, which they then eat; sabre-toothed blennies mimic other species of fish that specialize in cleaning large fish, and then take bites out of their clients' fins once they have been granted privileged access); avoiding being eaten (prey animals variously resemble tree bark, twigs, fresh green leaves, curled-up dead leaves, flowers, rose thorns, seaweed fronds, stones, bird droppings and other animals known to be poisonous or venomous); decoying predators away from young (avocets and many other ground-nesting birds mimic the attitude and gait of a bird with a broken wing); obtaining care of eggs (cuckoo eggs resemble the eggs of the particular host species parasitized; the females of certain species of mouthbreeder fish have dummy eggs painted on their flanks to attract males to take real eggs into their mouths and brood them).

In all cases, there is a temptation to think that the mimicry won't work unless it is perfect. In the particular case of the wasp orchid, I made much of the perceptual imperfections of wasps and other victims of mimicry. To my eyes, indeed, orchids are not all that uncanny in their resemblances to wasps, bees or flies. The resemblance of a leaf insect to a leaf is far more exact to my eyes, possibly because my eyes are more like the eyes of the predators (presumably birds) against which leaf mimicry is aimed.

But there is a more general sense in which it is wrong to suggest that mimicry has to be perfect if it is to work at all. However good the eyes of, say, a predator may be, the conditions for seeing are not always perfect. Moreover, there will almost inevitably be a continuum of seeing conditions, from very bad to very good. Think about some object you know really well, so well that you could never possibly mistake it for anything else. Or think of a person—say, a close friend, so dear and familiar that you could never mistake her for anybody else. But now imagine that she is walking toward you from a great distance. There must be a distance so great that you don't see her at all. And a distance so close that you see every feature, every eyelash, every pore. At intermediate distances, there is no sudden transformation. There is a gradual fade-in or fade-out of recognizability. Military manuals of riflemanship spell it out: "At two hundred yards, all parts of the body are distinctly seen. At three hundred yards, the outline of the face is blurred. At four hundred yards, no face. At six hundred yards, the head is a dot and the body tapers. Any questions?" In the case of the gradually approaching friend, admittedly you may suddenly recognize her. But in this case distance provides a gradient of *probability* of sudden recognition.

Distance, in one way or another, provides a gradient of visibility. It is essentially gradual. For any degree of resemblance between a model and a mimic, whether the resemblance is brilliant or almost nonexistent, there must be a distance at which a predator's eyes would be fooled and a slightly shorter distance at which they are less likely to be fooled. As evolution proceeds, resemblances of gradually improving perfection can therefore be favored by natural selection, in that the critical distance for being fooled gradually moves nearer. I use "predator's eyes" to stand for "the eyes of whoever needs to be fooled." In some cases it will be prey's eyes, foster-parent's eyes, female fish's eyes, and so on.

I have demonstrated this effect in public lectures to audiences of children. My colleague Dr. George McGavin, of the Oxford University Museum, kindly manufactured for me a model "woodland floor" strewn with twigs, dead leaves and moss. On it he artfully positioned dozens of dead insects. Some of these, such as a metallic-blue beetle, were quite conspicuous; others, including stick insects and leaf-mimicking butterflies, were exquisitely camouflaged; yet others, such as a brown cockroach, were intermediate. Children were invited out of the audience and asked to walk slowly toward the tableau, looking for insects and singing out as they spotted each one. When they were sufficiently far away, they couldn't see even the conspicuous insects. As they approached, they saw the conspicuous insects first, then those, like the cockroach, of intermediate visibility, and finally the well-camouflaged ones. The very best-camouflaged insects evaded detection even when the children were staring at them at close range, and the children gasped when I pointed them out.

Distance is not the only gradient about which one can make this kind of argument. Twilight is another. At dead of night, almost nothing can be seen, and even a very crude resemblance of mimic to model will pass muster. At high noon, only a meticulously accurate mimic may escape detection. Between these times, at daybreak and dusk, in the gloaming or just on a dull overcast day, in a fog or in a rainstorm, a smooth and unbroken continuum of visibilities obtains. Once again, resemblances of gradually increasing accuracy will be favored by natural selection, because for any given goodness of resemblance there will be a level of visibility at which that particular goodness of resemblance makes all the difference. As evolution proceeds, progressively improving resemblances confer survival advantage, because the critical light intensity for being fooled becomes gradually brighter.

A similar gradient is provided by angle of vision. An insect mimic, whether good or bad, will sometimes be seen out of the corner of a predator's eye. At other times it will be seen in merciless full-frontal aspect. There must be an angle of view so peripheral that the poorest possible mimic will escape detection. There must be a view so central that even the most brilliant mimic will be in danger. Between the two is a steady gradient of view, a continuum of angles. For any given level of perfection of mimicry, there will be a critical angle at which slight improvement or disimprovement makes all the difference. As evolution proceeds, resemblances of smoothly increasing quality are favored, because the critical angle for being fooled becomes gradually more central.

Quality of enemies' eyes and brains can be regarded as yet another gradient, and I have already hinted at it in earlier parts of this chapter. For any degree of resemblance between a model and a mimic, there is likely to be an eye that will be fooled and an eye that will not be fooled. Again, as evolution proceeds, resemblances of smoothly increasing quality are favored, because predator eyes of greater and greater sophistication are being fooled. I don't mean that the predators are evolving better eyes in parallel to the improving mimicry, though they might. I mean that there exist, somewhere out there, predators with good eyes and predators with poor eyes. All these predators constitute a danger. A poor mimic fools only the predators with poor eyes. A good mimic fools almost all the predators. There is a smooth continuum in between.

Mention of poor eyes and good eyes brings me to the creationist's favorite conundrum. What is the use of half an eye? How can natural selection favor an eye that is less than perfect? I have treated the question before and have laid out a spectrum of intermediate eyes, drawn from those that actually exist in the various phyla of the animal kingdom. Here I shall incorporate eyes in the rubric I have established of theoretical gradients. There is a gradient, a continuum, of tasks for which an eye might be used. I am at present using my eyes for recognizing letters of the alphabet as they appear on a computer screen. You need good, high-acuity eyes to do that. I have reached an age when I can no longer read without the aid of glasses, at present quite weakly magnifying ones. As I get older still, the strength of my prescription will steadily mount. Without my glasses, I shall find it gradually and steadily harder to see close detail. Here we have yet another continuum—a continuum of age.

Any normal human, however old, has better vision than an insect. There are tasks that can be usefully accomplished by people with relatively poor vision, all the way down to the nearly blind. You can play tennis with quite blurry vision, because a tennis ball is a large object, whose position and movement can be seen even if it is out of focus. Dragonflies' eyes, though poor by our standards, are good by insect standards, and dragonflies can hawk for insects on the wing, a task about as difficult as hitting a tennis ball. Much poorer eyes could be used for the task of avoiding crashing into a wall or walking over the edge of a cliff or into a river. Eyes that are even poorer could tell when a shadow, which might be a cloud but could also portend a predator, looms overhead. And eyes that are still poorer could serve to tell the difference between night and day, which is useful for, among other things, synchronizing breeding seasons and knowing when to go to sleep. There is a continuum of tasks to which an eye might be put, such that for any given quality of eye, from magnificent to terrible, there is a level of task at which a marginal improvement in vision would make all the difference. There is therefore no difficulty in understanding the gradual evolution of the eye, from primitive and crude beginnings, through a smooth continuum of intermediates, to the perfection we see in a hawk or in a young human.

Thus the creationist's question—"What is the use of half an eye?"—is a lightweight question, a doddle to answer. Half an eye is just 1 percent better than 49 percent of an eye, which is already better than 48 percent, and the difference is significant. A more ponderous show of weight seems to lie behind the inevitable supplementary: "Speaking as a physicist,* I cannot believe that there has been enough time for an organ as complicated as the eye to have evolved from nothing. Do you really think there has been enough time?" Both questions stem from the Argument from Personal Incredulity. Audiences nevertheless appreciate an answer, and I have usually

*I hope this does not give offense. In support of my point, I cite the following from *Science and Christian Belief,* by a distinguished physicist, the Reverend John Polkinghorne (1994, p. 16): "Someone like Richard Dawkins can present persuasive pictures of how the sifting and accumulation of small differences can produce large-scale developments, but, instinctively, a physical scientist would like to see an estimate, however rough, of how many steps would take us from a slightly light-sensitive cell to a fully formed insect eye, and of approximately the number of generations required for the necessary mutations to occur."

fallen back on the sheer magnitude of geological time. If one pace represents one century, the whole of Anno Domini time is telescoped into a cricket pitch. To reach the origin of multicellular animals on the same scale, you'd have to slog all the way from New York to San Francisco.

It now appears that the shattering enormity of geological time is a steamhammer to crack a peanut. Trudging from coast to coast dramatizes the time *available* for the evolution of the eye. But a recent study by a pair of Swedish scientists, Dan Nilsson and Susanne Pelger, suggests that a ludicrously small fraction of that time would have been plenty. When one says "the" eye, by the way, one implicitly means the vertebrate eye, but serviceable image-forming eyes have evolved between forty and sixty times, independently from scratch, in many different invertebrate groups. Among these forty-plus independent evolutions, at least nine distinct design principles have been discovered, including pinhole eyes, two kinds of camera-lens eyes, curved-reflector ("satellite dish") eyes, and several kinds of compound eyes. Nilsson and Pelger have concentrated on camera eyes with lenses, such as are well developed in vertebrates and octopuses.

How do you set about estimating the time required for a given amount of evolutionary change? We have to find a unit to measure the size of each evolutionary step, and it is sensible to express it as a percentage change in what is already there. Nilsson and Pelger used the number of successive changes of 1 percent as their unit for measuring changes of anatomical quantities. This is just a convenient unit—like the calorie, which is defined as the amount of energy needed to do a certain amount of work. It is easiest to use the 1 percent unit when the change is all in one dimension. In the unlikely event, for instance, that natural selection favored bird-of-paradise tails of ever-increasing length, how many steps would it take for the tail to evolve from one meter to one kilometer in length? A 1 percent increase in tail length would not be noticed by the casual bird-watcher. Nevertheless, it takes surprisingly few such steps to elongate the tail to one kilometer—fewer than seven hundred.

Elongating a tail from one meter to one kilometer is all very well (and all very absurd), but how do you place the evolution of an eye on the same scale? The problem is that in the case of the eye, lots of things have to go on in lots of different parts, in parallel. Nilsson and Pelger's task was to set up computer models of evolving eyes to answer two questions. The first is essentially the question we posed again and again in the past several pages, but they asked it more systematically, using a computer: Is there a smooth gradient of change, from flat skin to full camera eye, such that every intermediate is an improvement? (Unlike human designers, natural selection can't go downhill—not even if there is a tempting higher hill on the other side of the valley.) Second—the question with which we began this section—how long would the necessary quantity of evolutionary change take?

In their computer models, Nilsson and Pelger made no attempt to simulate the internal workings of cells. They started their story after the invention of a single light-sensitive cell—it does no harm to call it a photocell. It would be nice, in the future, to do another computer model, this time at the level of the inside of the cell, to show how the first living photocell came into being by step-by-step modification of an earlier, more general-purpose cell. But you have to start somewhere, and Nilsson and Pelger started after the invention of the photocell. They worked at the level of tissues: the level of stuff made of cells rather than the level of individual cells. Skin is a tissue, so is the lining of the intestine, so is muscle and liver. Tissues can change in various ways under the influence of random mutation. Sheets of tissue can become larger or smaller in area. They can become thicker or thinner. In the special case of transparent tissues like lens tissue, they can change the refractive index (the light-bending power) of local parts of the tissue.

The beauty of simulating an eye, as distinct from, say, the leg of a running cheetah, is that its efficiency can be easily measured, using the laws of elementary optics. The eye is represented as a two-dimensional cross section, and the computer can easily calculate its visual acuity, or spatial resolution, as a single real number. It would be much harder to come up with an equivalent numerical expression for the efficacy of a cheetah's leg or backbone. Nilsson and Pelger began with a flat retina atop a flat pigment layer and surmounted by a flat, protective transparent layer. The transparent layer was allowed to undergo localized random mutations of its refractive index. They then let the model deform itself at random, constrained only by the requirement that any change must be small and must be an improvement on what went before.

The results were swift and decisive. A trajectory of steadily mounting acuity led unhesitatingly from the flat beginning through a shallow indentation to a steadily deepening cup, as the shape of the model eye deformed itself on the computer screen. The transparent layer thickened to fill the cup and smoothly bulged its outer surface in a curve. And then, almost like a conjuring trick, a portion of this transparent filling condensed into a local, spherical subregion of higher refractive index. Not uniformly higher, but a gradient of refractive index such that the spherical region functioned as an excellent graded-index lens. Graded-index lenses are unfamiliar to human lensmakers but they are common in living eyes. Humans make lenses by grinding glass to a particular shape. We make a compound lens, like the expensive violet-tinted lenses of modern cameras, by mounting several lenses together, but each one of those individual lenses is made of uniform glass through its whole thickness. A graded-index lens, by contrast, has a continuously varying refractive index within its own substance. Typically, it has a high refractive index near the center of the lens. Fish eyes have graded-index lenses. Now it has long been known that, for a graded-index lens, the most aberration-free results are obtained when you achieve a particular theoretical optimum value for the ratio between the focal length of the lens and the radius. This ratio is called Mattiessen's ratio. Nilsson and Pelger's computer model homed in unerringly on Mattiessen's ratio.

And so to the question of how long all this evolutionary change might have taken. In order to answer this, Nilsson and Pelger had to make some assumptions about genetics in natural populations. They needed to feed their model plausible values of quantities such as "heritability." Heritability is a measure of how far variation is governed by heredity. The favored way of measuring it is to see how much monozygotic (that is, "identical") twins resemble each other compared with ordinary twins. One study found the heritability of leg length in male humans to be 77 percent. A heritability of 100 percent would mean that you could measure one identical twin's leg to obtain perfect knowledge of the other twin's leg length, even if the twins were reared apart. A heritability of 0 percent would mean that the legs of monozygotic twins are no more similar to each other than to the legs of random members of a specified population in a given environment. Some other heritabilities measured for humans are 95 percent for head breadth, 85 percent for sitting height, 80 percent for arm length and 79 percent for stature.

Heritabilities are frequently more than 50 percent, and Nilsson and Pelger therefore felt safe in plugging a heritability of 50 percent into their eye model. This was a conservative, or "pessimistic," assumption. Compared with a more realistic assumption of, say, 70 percent, a pessimistic assumption tends to increase their final estimate of the time taken for the eye to evolve. They wanted to err on the side of overestimation because we are intuitively skeptical of short estimates of the time taken to evolve something as complicated as an eye.

For the same reason, they chose pessimistic values for the coefficient of variation (that is, for how much variation there typically is in the population) and the intensity of selection (the amount of survival advantage improved eyesight confers). They even went so far as to assume that any new generation differed in only one part of the eye at a time: simultaneous changes in different parts of the eye, which would have greatly speeded up evolution, were outlawed. But even with these conservative assumptions, the time taken to evolve a fish eye from flat skin was minuscule: fewer than four hundred thousand generations. For the kinds of small animals we are talking about, we can assume one generation per year, so it seems that it would take less than half a million years to evolve a good camera eye.

In the light of Nilsson and Pelger's results, it is no wonder "the" eye has evolved at least forty times independently around the animal kingdom. There has been enough time for it to evolve from scratch fifteen hundred times in succession within any one lineage. Assuming typical generation lengths for small animals, the time needed for the evolution of the eye, far from stretching credulity with its vastness, turns out to be too short for geologists to measure! It is a geological blink.

Do good by stealth. A key feature of evolution is its gradualness. This is a matter of principle rather than fact. It may or may not be the case that some episodes of evolution take a sudden turn. There may be punctuations of rapid evolution, or even abrupt macromutations—major changes dividing a child from both its parents. There certainly are sudden extinctions—perhaps caused by great natural catastrophes such as comets striking the earth—and these leave vacuums to be filled by rapidly improving understudies, as the mammals replaced the dinosaurs. Evolution is very possibly not, in actual fact, always gradual. But it must be gradual when it is being used to explain the coming into existence of complicated, apparently designed objects, like eyes. For if it is not gradual in these cases, it ceases to have any explanatory power at all. Without gradualness in these cases, we are back to miracle, which is simply a synonym for the total absence of explanation.

The reason eyes and wasp-pollinated orchids impress us so is that they are improbable. The odds against their spontaneously assembling by luck are odds too great to be borne in the real world. Gradual evolution by small steps, each step being lucky but not *too* lucky, is the solution to the riddle. But if it is not gradual, it is no solution to the riddle: it is just a restatement of the riddle.

There will be times when it is hard to think of what the gradual intermediates may have been. These will be challenges to our ingenuity, but if our ingenuity fails, so much the worse for our ingenuity. It does not constitute evidence that there were no gradual intermediates. One of the most difficult challenges to our ingenuity in thinking of gradual intermediates is provided by the celebrated "dance language" of the bees, discovered in the classic work for which Karl von Frisch is best known. Here the end product of the evolution seems so complicated, so ingenious and far removed from anything we would ordinarily expect an insect to do, that it is hard to imagine the intermediates.

Honeybees tell each other the whereabouts of flowers by means of a carefully coded dance. If the food is very close to the hive, they do the "round dance." This just excites other bees, and they rush out and search in the vicinity of the hive. Not particularly remarkable. But *very* remarkable is what happens when the food is farther away from the hive. The forager who has discovered the food performs the so-called "waggle dance," and its form and timing tell the other bees both the compass direction and the distance from the hive of the food. The waggle dance is performed inside the hive on the vertical surface of the comb. It is dark in the hive, so the waggle dance is not seen by the other bees. It is felt by them, and also heard, for the dancing bee accompanies her performance with little rhythmic piping noises. The dance has the form of a figure eight, with a straight run in the middle. It is the direction of the straight run that, in the form of a cunning code, tells the direction of the food.

The straight run of the dance does not point directly toward the food. It cannot, since the dance is performed on the vertical surface of the comb and the alignment of the comb itself is fixed regardless of where the food might be. The food has to be located in horizontal geography. The vertical comb is more like a map pinned to the wall. A line drawn on a wall map doesn't point directly toward a particular destination, but you can read the direction by means of an arbitrary convention.

To understand the convention the bees use, you must first know that bees, like many insects, navigate using the sun as a compass. We do this too, in an approximate way. The method has two drawbacks. First, the sun is often hidden behind clouds. Bees solve this problem by means of a sense we don't have. Again, it was von Frisch who discovered that they can see the direction of polarization of light and this tells them where the sun is even if the sun itself is invisible. The second problem with a sun compass is that the sun "moves" across the sky as the hours progress. Bees cope with this by using an internal clock. Von Frisch found, almost unbelievably, that dancing bees trapped in the hive for hours after their foraging expedition would slowly rotate the direction of the straight run of the dance, as if this run were the hour hand of a twenty-four-hour clock. They could not see the sun inside the hive, but they were slowly angling the direction of their dance in order to keep pace with the movement of the sun, which, their internal clocks told them, must be going on outside. Fascinatingly, bee races hailing from the Southern Hemisphere do the same thing in reverse, just as they should.

Now, to the dance code itself. A dance run pointing straight up the comb signals that food is in the same direction as the sun. Straight down signals food in the exact opposite direction. All intermediate angles signal what you would expect. Fifty degrees to the left of the vertical signifies 50° to the left of the sun's direction in the horizontal plane. The accuracy of the dance is not to the nearest degree, however. Why should it be, for it is our arbitrary convention to divide the compass into 360°? Bees divide the compass into about 8 bee degrees. Actually, this is approximately what we do when we are not professional navigators. We divide our informal compass into eight quadrants: N, NE, E, SE, S, SW, W, NW.

The bee dance codes the distance of food, too. Or rather, various aspects of the dance—the rate of turning, the rate of waggling, the rate of peeping—are correlated with the distance of the food, and any one of them or any combination of them could therefore be used by the other bees to read the distance. The nearer the food, the faster the dance. You can remember this by reflecting that a bee that has found food near the hive might be expected to be more excited, and less tired, than a bee that has found food a long distance away. This is more than just an *aide-memoire*; it gives a clue to how the dance evolved, as we shall see.

To summarize, a foraging bee finds a good source of food. She returns to the hive, laden with nectar and pollen, and delivers her cargo to receiving workers. Then she begins her dance. Somewhere on a vertical comb, it doesn't matter where, she rushes round and round in a tight figure eight. Other worker bees cluster around her, feeling and listening. They count the rate of peeping and perhaps the rate of turning too. They measure, relative to the vertical, the angle of the straight run of the dance while the dancer is waggling her abdomen. They then move to the hive's door and burst out of the darkness into the sunlight. They observe the position of the sun—not its vertical height but its compass bearing in the horizontal plane. And they fly off in a straight line, whose angle relative to the sun matches the angle of the original forager's dance relative to the vertical on the comb. They keep flying on this bearing, not for an indefinite distance but for a distance (inversely) proportional to (the logarithm of) the rate of peeping of the original dancer. Intriguingly, if the original bee had flown in a detour in order to find the food, she points her dance not in the direction of her detour but in the reconstructed compass direction of the food.

The story of the dancing bees is hard to believe, and some have disbelieved it. I'll return to the skeptics, and to the recent experiments that finally clinched the evidence, in the next chapter. In this chapter, I want to discuss the gradual evolution of the bee dance. What might the intermediate stages in its evolution have looked like, and how did they work when the dance was still incomplete?

The way the question is phrased is not quite right, by the way. No creature makes a living at being an "incomplete," "intermediate stage." The ancient, long-dead bees whose dances can be interpreted, with hindsight, as intermediates on the way to the modern honeybee dance made a good living. They lived full bee lives and had no thought of being "on the way" to something "better." Moreover, our "modern" bee dance may not be the last word but may evolve into something even more spectacular when we and our bees are gone. Nevertheless, we do have the puzzle of how the current bee dance could have evolved by gradual degrees. How might those graduated intermediates have looked, and how did they work?

Von Frisch himself has attended to the question, and he tackled it by looking around the family tree, at modern distant cousins of the honeybee. These are not ancestors of the honeybee, for they are its contemporaries. But they may retain features of the ancestors. The honeybee itself is a temperate-zone insect that nests for shelter in hollow trees or caves. Its closest relatives are tropical bees that can nest in the open, hanging their combs from tree boughs or rocky outcrops. They are therefore able to see the sun while dancing, and they don't have to resort to the convention of letting the vertical "stand for" the direction of the sun. The sun can stand for itself.

One of these tropical relations, the dwarf bee *Apis florea,* dances on the horizontal surface on top of the comb. The straight run of the dance points directly toward the food. There is no need for a map convention; direct pointing will do. A plausible transitional stage on the road to the honeybee, certainly, but we still have to think about the other intermediates that preceded and followed this stage. What could have been the forerunners of the dwarf bee's dance? Why should a bee that has recently found food rush round and round in a figure eight whose straight run points toward the food? The suggestion is that it is a ritualized form of the take-off run. Before the dance evolved, von Frisch suggests, a forager that has just unloaded food would simply take off in the same direction, to fly back to the food source. Preparatory to launching itself into the air, it would turn its face in the right direction and might walk a few steps. Natural selection would have favored any tendency to exaggerate or prolong the take-off run if it encouraged other bees to follow. Perhaps the dance is a kind of ritually repeated take-off run. This is plausible because, whether or not they use a dance, bees frequently use the more direct tactic of simply following each other to food sources. Another fact that gives the idea plausibility is that dancing bees hold their wings out slightly, as though preparing to fly, and they vibrate the wing muscles, not vigorously enough to take off but enough to make the noise that is an important part of the dance signal.

An obvious way to prolong and exaggerate the take-off run is to repeat it. Repeating it means going back to the start and again taking a few steps in the direction of the food. There are two ways of going back to the start: you can turn right or turn left at the end of the runway. If you consistently turn left or consistently turn right, it will be ambiguous which direction is the true take-off direction and which the return journey to the start of the runway. The best way to remove ambiguity is to turn alternately left and right. Hence the natural selection of the figure-eight pattern.

But how did the relationship between distance of food and rate of dancing evolve? If the rate of dancing were positively related to the distance of the food, it would be hard to explain. But, you'll remember, it is actually the other way around: the closer the food, the faster the dance. This immediately suggests a plausible pathway of gradual evolution. Before the dance proper evolved, foragers might have performed their ritualized repetition of the take-off run but at no particular speed. The rate of dancing would have been whatever they happened to feel like. Now, if you had just flown home from several miles away, laden to the gunwales with nectar and pollen, would you feel like charging at high speed around the comb? No, you would probably be exhausted. On the other hand, if you had just discovered a rich source of food rather close to the hive, your short homeward journey would have left you fresh and energetic. It is not difficult to imagine how an original accidental relationship between distance of food and slowness of dance could have become ritualized into a formal, reliable code.

Now for the most challenging intermediate of all. How did an ancient dance in which the straight run pointed directly toward the food get transformed into a dance in which the angle relative to the vertical becomes a code for the angle of the food relative to the sun? Such a transformation was necessary partly because the interior of the honeybee hive is dark and you can't see the sun, and partly because when dancing on a vertical comb you can't point directly toward the food unless the surface itself happens to be pointing toward the food. But it isn't enough to show that some such transformation was necessary. We also have to explain how this difficult transition was achieved via a plausible series of step-by-step intermediates.

It seems baffling, but a singular fact about the insect nervous system comes to our rescue. The following remarkable experiment has been done on a variety of insects, from beetles to ants. Start with a beetle walking along a horizontal wooden board in the presence of an electric light. The first thing to show is that the insect is using a light-compass. Shift the position of the lightbulb, and the insect will change its direction accordingly. If it was holding a bearing of, say, 30° to the light, it will change its path so as to maintain a bearing of 30° to the new position of the light. In fact you can steer the beetle wherever you like, using the light beam as a tiller. This fact about insects has long been known: they use the sun (or moon or stars) as a compass, and you can easily fool them with a lightbulb. So far so good. Now for the interesting experiment. Switch the light off and at the same moment tilt the board into the vertical. Undaunted, the beetle continues walking. And, *mirabile dictu,* it shifts its direction of walking so that its angle relative to the vertical is the same as the previous angle relative to the light: 30° in our example. Nobody knows why this happens, but it does. It seems to betray an accidental quirk of the insect nervous system—a confusion of senses, a crossing of wires between the gravity sense and the sense of sight, perhaps a bit like our seeing a flash of light when hit on the head. At all events, it probably provided the necessary bridge for the evolution of the "vertical stands for sun" code of the honeybee's dance.

Revealingly, if you switch on a light inside a hive, honeybees abandon their gravity sense and use the direction of the light to stand, directly, for the sun in their code. This fact, long known, has been exploited in one of the most ingenious experiments ever performed, the experiment that finally clinched the evidence that the honeybee dance really works. I'll return to this in the next chapter. Meanwhile, we have found a plausible series of graded intermediates by which the modern bee dance could have evolved from simpler beginnings. The story as I have told it, based on von Frisch's ideas, may not actually be the right one. But something a bit like it surely did happen. I told the story as an answer to the natural skepticism—the Argument from Personal Incredulity—that arises in people when they are faced with a really ingenious or complicated natural phenomenon. The skeptic says, "I cannot imagine a plausible series of intermediates, therefore there were none, and the phenomenon arose by a spontaneous miracle." Von Frisch has provided a plausible series of intermediates. Even if it is not quite the right series, the fact that it is plausible is enough to confound the Argument from Personal Incredulity. The same is true of all the other examples we have looked at, from wasp-mimicking orchids to camera eyes.

Any number of curious and intriguing facts of nature could be mustered by people skeptical of gradualistic Darwinism. I have been asked to explain, for example, the gradual evolution of those creatures that live in the deep trenches of the Pacific Ocean, where there is no light and where water pressures may exceed 1000 atmospheres. A whole community of animals has grown up around hot, volcanic vents deep in the Pacific trenches. A whole alternative biochemistry is run by bacteria, using the heat from the vents and metabolizing sulfur instead of oxygen. The community of larger animals is ultimately dependent on these sulfur bacteria, just as ordinary life is dependent on green plants capturing energy from the sun.

The animals in the sulfur community are all relatives of more conventional animals found elsewhere. How did they evolve and through what intermediate stages? Well, the form of the argument will be exactly the same. All we need for our explanation is at least one natural gradient, and gradients abound as we descend in the sea. A thousand atmospheres is a horrendous pressure, but it is only quantitatively greater than 999 atmospheres, which is only quantitatively greater than 998 and so on. The sea bottom offers gradients of depth from 0 feet through all intermediates to 33,000 feet. Pressures vary smoothly from 1 atmosphere to 1000 atmospheres. Light intensities vary smoothly from bright daylight near the surface to total darkness in the deeps, relieved only by rare clusters of luminescent bacteria in the luminous organs of fishes. There are no sharp cut-offs. For every level of pressure and darkness already adapted to, there will be a design of animal, only slightly different from existing animals, that can survive one fathom deeper, one lumen darker. For every . . . but this chapter is more than long enough. You know my methods, Watson. Apply them.

CHAPTER 4

GOD'S UTILITY FUNCTION

My clerical correspondent of the previous chapter found faith through a wasp. Charles Darwin lost his with the help of another: "I cannot persuade myself," Darwin wrote, "that a beneficent and omnipotent God would have designedly created the Ichneumonidae with the express intention of their feeding within the living bodies of Caterpillars." Actually Darwin's gradual loss of faith, which he downplayed for fear of upsetting his devout wife Emma, had more complex causes. His reference to the Ichneumonidae was aphoristic. The macabre habits to which he referred are shared by their cousins the digger wasps, whom we met in the previous chapter. A female digger wasp not only lays her egg in a caterpillar (or grasshopper or bee) so that her larva can feed on it but, according to Fabre and others, she carefully guides her sting into each ganglion of the prey's central nervous system, so as to paralyze it *but not kill it.* This way, the meat keeps fresh. It is not known whether the paralysis acts as a general anesthetic, or if it is like curare in just freezing the victim's ability to move. If the latter, the prey might be aware of being eaten alive from inside but unable to move a muscle to do anything about it. This sounds savagely cruel but as we shall see, nature is not cruel, only pitilessly indifferent. This is one of the hardest lessons for humans to learn. We cannot admit that things might be neither good nor evil, neither cruel nor kind, but simply callous—indifferent to all suffering, lacking all purpose.

We humans have purpose on the brain. We find it hard to look at anything without wondering what it is "for," what the motive for it is, or the purpose behind it. When the obsession with purpose becomes pathological it is called paranoia—reading malevolent purpose into what is actually random bad luck. But this is just an exaggerated form of a nearly universal delusion. Show us almost any object or process, and it is hard for us to resist the "Why" question—the "What is it for?" question.

The desire to see purpose everywhere is a natural one in an animal that lives surrounded by machines, works of art, tools and other designed artifacts; an animal, moreover, whose waking thoughts are dominated by its own personal goals. A car, a tin opener, a screwdriver and a pitchfork all legitimately warrant the "What is it for?" question. Our pagan forebears would have asked the same question about thunder, eclipses, rocks and streams. Today we pride ourselves on having shaken off such primitive animism. If a rock in a stream happens to serve as a convenient stepping-stone, we regard its usefulness as an accidental bonus, not a true purpose. But the old temptation comes back with a vengeance when tragedy strikes—indeed, the very word "strikes" is an animistic echo: "Why, oh why, did the cancer/earthquake/hurricane have to strike *my* child?" And the same temptation is often positively relished when the topic is the origin of all things or the fundamental laws of physics, culminating in the vacuous existential question "Why is there something rather than nothing?"

I have lost count of the number of times a member of the audience has stood up after a public lecture I have given and said something like the following: "You scientists are very good at answering 'How' questions. But you must admit you're powerless when it comes to 'Why' questions." Prince Philip, Duke of Edinburgh, made this very point when he was in an audience at Windsor addressed by my colleague Dr. Peter Atkins. Behind the question there is always an unspoken but never justified implication that since science is unable to answer "Why" questions, there must be some other discipline that *is* qualified to answer them. This implication is, of course, quite illogical.

I'm afraid that Dr. Atkins gave the Royal Why fairly short shrift. The mere fact that it is possible to frame a question does not make it legitimate or sensible to do so. There are many things about which you can ask, "What is its temperature?" or "What color is it?" but you may not ask the temperature question or the color question of, say, jealousy or prayer. Similarly, you are right to ask the "Why" question of a bicycle's mudguards or the Kariba Dam, but at the very least you have no right to *assume* that the "Why" question deserves an answer when posed about a boulder, a misfortune, Mt. Everest or the universe. Questions can be simply inappropriate, however heartfelt their framing.

Somewhere between windscreen wipers and tin openers on the one hand and rocks and the universe on the other lie living creatures. Living bodies and their organs are objects that, unlike rocks, seem to have purpose written all over them. Notoriously, of course, the apparent purposefulness of living bodies has dominated the classic Argument from Design, invoked by theologians from Aquinas to William Paley to modern "scientific" creationists.

The true process that has endowed wings and eyes, beaks, nesting instincts and everything else about life with the strong illusion of purposeful design is now well understood. It is Darwinian natural selection. Our understanding of this has come astonishingly recently, in the last century and a half. Before Darwin, even educated people who had abandoned "Why" questions for rocks, streams and eclipses still implicitly accepted the legitimacy of the "Why" question where living creatures were concerned. Now only the scientifically illiterate do. But "only" conceals the unpalatable truth that we are still talking about an absolute majority.

Actually, Darwinians do frame a kind of "Why" question about living things, but they do so in a special, metaphorical sense. Why do birds sing, and what are wings for? Such questions would be accepted as a shorthand by modern Darwinians and would be given sensible answers in terms of the natural selection of bird ancestors. The illusion of purpose is so powerful that biologists themselves use the assumption of good design as a working tool. As we saw in the previous chapter, long before his epoch-making work on the bee dance Karl von Frisch discovered, in the teeth of strong orthodox opinion to the contrary, that some insects have true color vision. His clinching experiments were stimulated by the simple observation that bee-pollinated flowers go to great trouble to manufacture colored pigments. Why would they do this if bees were color-blind? The metaphor of purpose—more precisely, the assumption that Darwinian selection is involved—is here being used to make a strong inference about the world. It would have been quite wrong for von Frisch to have said, "Flowers are colored, therefore bees must have color vision." But it was right for him to say, as he did, "Flowers are colored, therefore it is at least worth my while working hard at some new experiments to test the hypothesis that they have color vision." What he found when he looked into the matter in detail was that bees have good color vision but the spectrum they see is shifted relative to ours. They can't see red light (they might give the name "infra yellow" to what we call red). But they can see into the range of shorter wavelengths we call ultraviolet, and they see ultraviolet as a distinct color, sometimes called "bee purple."

When he realized that bees see in the ultraviolet part of the spectrum, von Frisch again did some reasoning using the metaphor of purpose. What, he asked himself, do bees use their ultraviolet sense for? His thoughts returned full circle—to flowers. Although we can't see ultraviolet light, we can make photographic film that is sensitive to it, and we can make filters that are transparent to ultraviolet light but cut out "visible" light. Acting on his hunch, von Frisch took some ultraviolet photographs of flowers. To his delight, he saw patterns of spots and stripes that no human eye had ever seen before. Flowers that to us look white or yellow are in fact decorated with ultraviolet patterns, which often serve as runway markers to guide the bees to the nectaries. The assumption of apparent purpose had paid off once again: flowers, if they were well designed, would exploit the fact that bees can see ultraviolet wavelengths.

When he was an old man, von Frisch's most famous work—on the dance of the bees, which we discussed in the last chapter—was called into question by an American biologist named Adrian Wenner. Fortunately, von Frisch lived long enough to see his work vindicated by another American, James L. Gould, now at Princeton, in one of the most brilliantly conceived experiments of all biology. I'll briefly tell the story, because it is relevant to my point about the power of the "as if designed" assumption.

Wenner and his colleagues did not deny that the dance happens. They did not even deny that it contains all the information von Frisch said it did. What they did deny is that other bees read the dance. Yes, Wenner said, it is true that the direction of the straight run of the waggle dance relative to the vertical is related to the direction of food relative to the sun. But no, other bees don't receive this information from the dance. Yes, it is true that the rates of various things in the dance can be read as information about the distance of food. But there is no good evidence that the other bees read the information. They could be ignoring it. Von Frisch's evidence, the skeptics said, was flawed, and when they repeated his experiments with proper "controls" (that is, by taking care of alternative means by which bees might find food), the experiments no longer supported von Frisch's dance-language hypothesis.

This was where Jim Gould came into the story with his exquisitely ingenious experiments. Gould exploited a long-known fact about honeybees, which you will remember from the previous chapter. Although they usually dance in the dark, using the straight-up direction in the vertical plane as a coded token of the sun's direction in the horizontal plane, they will effortlessly switch to a possibly more ancestral way of doing things if you turn on a light inside the hive. They then forget all about gravity and use the lightbulb as their token sun, allowing it to determine the angle of the dance directly. Fortunately, no misunderstandings arise when the dancer switches her allegiance from gravity to the lightbulb. The other bees "reading" the dance switch their allegiance in the same way, so the dance still carries the same meaning: the other bees still head off looking for food in the direction the dancer intended.

Now for Jim Gould's masterstroke. He painted a dancing bee's eyes over with black shellac, so that she couldn't see the lightbulb. She therefore danced using the normal gravity convention. But the other bees following her dance, not being blindfolded, could see the lightbulb. They interpreted the dance as if the gravity convention had been dropped and replaced by the lightbulb "sun" convention. The dance followers measured the angle of the dance relative to the light, whereas the dancer herself was aligning it relative to gravity. Gould was, in effect, forcing the dancing bee to lie about the direction of the food. Not just lie in a general sense, but lie in a particular direction that Gould could precisely manipulate. He did the experiment not with just one blindfolded bee, of course, but with a proper statistical sample of bees and variously manipulated angles. And it worked. Von Frisch's original dance-language hypothesis was triumphantly vindicated.

I didn't tell this story for fun. I wanted to make a point about the negative as well as the positive aspects of the assumption of good design. When I first read the skeptical papers of Wenner and his colleagues, I was openly derisive. And this was not a good thing to be, even though Wenner eventually turned out to be wrong. My derision was based entirely on the "good design" assumption. Wenner was not, after all, denying that the dance happened, nor that it embodied all the information von Frisch had claimed about the distance and direction of food. Wenner simply denied that the other bees read the information. And this was too much for me and many other Darwinian biologists to stomach. The dance was so complicated, so richly contrived, so finely tuned to its apparent purpose of informing other bees of the distance and direction of food. This fine tuning could not have come about, in our view, other than by natural selection. In a way, we fell into the same trap as creationists do when they contemplate the wonders of life. The dance simply had to be doing something useful, and this presumably meant helping foragers to find food. Moreover, those very aspects of the dance that were so finely tuned—the relationship of its angle and speed to the direction and distance of food—had to be doing something useful too. Therefore, in our view, Wenner just had to be wrong. So confident was I that, even if I had been ingenious enough to think of Gould's blindfold experiment (which I certainly wasn't), I would not have bothered to do it.

Gould not only was ingenious enough to think of the experiment but he also bothered to do it, because he was not seduced by the good-design assumption. It is a fine tightrope we are walking, however, because I suspect that Gould—like von Frisch before him, in his color research—had enough of the good-design assumption in his head to believe that his remarkable experiment had a respectable chance of success and was therefore worth spending time and effort on.

I now want to introduce two technical terms, "reverse engineering" and "utility function." In this section, I am influenced by Daniel Dennett's superb book *Darwin's Dangerous Idea*. Reverse engineering is a technique of reasoning that works like this. You are an engineer, confronted with an artifact you have found and don't understand. You make the working assumption that it was designed for some purpose. You dissect and analyze the object with a view to working out what problem it would be good at solving: "If I had wanted to make a machine to do so-and-so, would I have made it like this? Or is the object better explained as a machine designed to do such-and-such?"

The slide rule, talisman until recently of the honorable profession of engineer, is in the electronic age as obsolete as any Bronze Age relic. An archaeologist of the future, finding a slide rule and wondering about it, might note that it is handy for drawing straight lines or for buttering bread. But to assume that either of these was its original purpose violates the economy assumption. A mere straight-edge or butter knife would not have needed a sliding member in the middle of the rule. Moreover, if you examine the spacing of the graticules you find precise logarithmic scales, too meticulously disposed to be accidental. It would dawn on the archaeologist that, in an age before electronic calculators, this pattern would constitute an ingenious trick for rapid multiplication and division. The mystery of the slide rule would be solved by reverse engineering, employing the assumption of intelligent and economical design.

"Utility function" is a technical term not of engineers but of economists. It means "that which is maximized." Economic planners and social engineers are rather like architects and real engineers in that they strive to maximize something. Utilitarians strive to maximize "the greatest happiness for the greatest number" (a phrase that sounds more intelligent than it is, by the way). Under this umbrella, the utilitarian may give long-term stability more or less priority at the expense of short-term happiness, and utilitarians differ over whether they measure "happiness" by monetary wealth, job satisfaction, cultural fulfillment or personal relationships. Others avowedly maximize their own happiness at the expense of the common welfare, and they may dignify their egoism by a philosophy that states that general happiness will be maximized if one takes care of oneself. By watching the behavior of individuals throughout their lives, you should be able to reverse-engineer their utility functions. If you reverse-engineer the behavior of a country's government, you may conclude that what is being maximized is employment and universal welfare. For another country, the utility function may turn out to be the continued power of the president, or the wealth of a particular ruling family, the size of the sultan's harem, the stability of the Middle East or maintaining the price of oil. The point is that more than one utility function can be imagined. It isn't always obvious what individuals, or firms, or governments are striving to maximize. But it is probably safe to assume that they are maximizing something. This is because *Homo sapiens* is a deeply purpose-ridden species. The principle holds good even if the utility function turns out to be a weighted sum or some other complicated function of many inputs.

Let us return to living bodies and try to extract their utility function. There could be many but, revealingly, it will eventually turn out that they all reduce to one. A good way to dramatize our task is to imagine that living creatures were made by a Divine Engineer and try to work out, by reverse engineering, what the Engineer was trying to maximize: What was God's Utility Function?

Cheetahs give every indication of being superbly designed for something, and it should be easy enough to reverse-engineer them and work out their utility function. They appear to be well designed to kill antelopes. The teeth, claws, eyes, nose, leg muscles, backbone and brain of a cheetah are all precisely what we should expect if God's purpose in designing cheetahs was to maximize deaths among antelopes. Conversely, if we reverse-engineer an antelope we find equally impressive evidence of design for precisely the opposite end: the survival of antelopes and starvation among cheetahs. It is as though cheetahs had been designed by one deity and antelopes by a rival deity. Alternatively, if there is only one Creator who made the tiger and the lamb, the cheetah and the gazelle, what is He playing at? Is He a sadist who enjoys spectator blood sports? Is He trying to avoid overpopulation in the mammals of Africa? Is He maneuvering to maximize David Attenborough's television ratings? These are all intelligible utility functions that might have turned out to be true. In fact, of course, they are all completely wrong. We now understand the single Utility Function of life in great detail, and it is nothing like any of those.

Chapter 1 will have prepared the reader for the view that the true utility function of life, that which is being maximized in the natural world, is DNA survival. But DNA is not floating free; it is locked up in living bodies and it has to make the most of the levers of power at its disposal. DNA sequences that find themselves in cheetah bodies maximize their survival by causing those bodies to kill gazelles. Sequences that find themselves in gazelle bodies maximize their survival by promoting opposite ends. But it is DNA survival that is being maximized in both cases. In this chapter, I am going to do a reverse-engineering job on a number of practical examples and show how everything makes sense once you assume that DNA survival is what is being maximized.

The sex ratio—the proportion of males to females—in wild populations is usually 50:50. This seems to make no economic sense in those many species in which a minority of males has an unfair monopoly of the females: the harem system. In one well-studied population of elephant seals, 4 percent of the males accounted for 88 percent of all the copulations. Never mind that God's Utility Function in this case seems so unfair for the bachelor majority. What is worse, a cost-cutting, efficiency-minded deity would be bound to spot that the deprived 96 percent are consuming half the population's food resources (actually more than half, because adult male elephant seals are much bigger than females). The surplus bachelors do nothing except wait for an opportunity to displace one of the lucky 4 percent of harem masters. How can the existence of these unconscionable bachelor herds possibly be justified? Any utility function that paid even a little attention to the economic efficiency of the community would dispense with the bachelors. Instead, there would be just enough males born to fertilize the females. This apparent anomaly, again, is explained with elegant simplicity once you understand the true Darwinian Utility Function: maximizing DNA survival.

I'll go into the example of the sex ratio in a little detail, because its utility function lends itself subtly to an economic treatment. Charles Darwin confessed himself baffled: "I formerly thought that when a tendency to produce the two sexes in equal numbers was advantageous to the species, it would follow from natural selection, but I now see that the whole problem is so intricate that it is safer to leave its solution for the future." As so often, it was the great Sir Ronald Fisher who stood up in Darwin's future. Fisher reasoned as follows.

All individuals born have exactly one mother and one father. Therefore the total reproductive success, measured in distant descendants, of all males alive must equal that of all females alive. I don't mean of *each* male and female, because some individuals clearly, and importantly, have more success than others. I am talking about the totality of males compared with the totality of females. This total posterity must be divided up between the individual males and females—not divided equally, but divided. The reproductive cake that must be divided among all males is equal to the cake that must be divided among all females. Therefore if there are, say, more males than females in the population, the average slice of cake per male must be smaller than the average slice of cake per female. It follows that the average reproductive success (that is, the expected number of descendants) of a male compared with the average reproductive success of a female is solely determined by the male-female ratio. An average member of the minority sex has a greater reproductive success than an average member of the majority sex. Only if the sex ratio is even and there is no minority will the sexes enjoy equal reproductive success. This remarkably simple conclusion is a consequence of pure armchair logic. It doesn't depend on any empirical facts at all, except the fundamental fact that all children born have one father and one mother.

Sex is usually determined at conception, so we may assume that an individual has no power to determine his or her (for once the circumlocution is not ritual but necessary) sex. We shall assume, with Fisher, that a parent might have power to determine the sex of its offspring. By "power," of course, we don't mean power consciously or deliberately wielded. But a mother might have a genetic predisposition to generate a vaginal chemistry slightly hostile to son-producing but not to daughter-producing sperms. Or a father might have a genetic tendency to manufacture more daughter-producing sperms than son-producing sperms. However it might in practice be done, imagine yourself as a parent trying to decide whether to have a son or a daughter. Again, we are not talking about conscious decisions but about the selection of generations of genes acting on bodies to influence the sex of their offspring.

If you were trying to maximize the number of your grandchildren, should you have a son or a daughter? We have already seen that you should have a child of whichever sex is in the minority in the population. That way, your child can expect a relatively large share of reproductive activity and you can expect a relatively large number of grandchildren. If neither sex is rarer than the other—if, in other words, the ratio is already 50:50—you cannot benefit by preferring one sex or the other. It doesn't matter whether you have a son or a daughter. A 50:50 sex ratio is therefore referred to as evolutionarily stable, using the term coined by the great British evolutionist John Maynard Smith. Only if the existing sex ratio is something other than 50:50 does a bias in your choice pay. As for the question of why individuals should try to maximize their grandchildren and later descendants, it will hardly need asking. Genes that cause individuals to maximize their descendants are the genes we expect to see in the world. The animals we are looking at inherit the genes of successful ancestors.

It is tempting to express Fisher's theory by saying that 50:50 is the "optimum" sex ratio, but this is strictly incorrect. The optimum sex to choose for a child is male if males are in a minority, female if females are in a minority. If neither sex is in a minority, there is no optimum: the well-designed parent is strictly indifferent about whether a son or a daughter will be born. Fifty-fifty is said to be the evolutionarily stable sex ratio because natural selection does not favor any tendency to deviate from it, and if there is any deviation from it natural selection favors a tendency to redress the balance.

Moreover, Fisher realized that it isn't strictly the numbers of males and females that are held at 50:50 by natural selection, but what he called the "parental expenditure" on sons and daughters. Parental expenditure means all the hard-won food poured into the mouth of a child; and all the time and energy spent looking after it, which could have been spent doing something else, such as looking after another child. Suppose, for instance, that parents in a particular seal species typically spend twice as much time and energy on rearing a son as on rearing a daughter. Bull seals are so massive compared with cows that it is easy to believe (though probably inaccurate in fact) that this might be the case. Think what it would mean. The true choice open to the parent is not "Should I have a son or a daughter?" but "Should I have a son or *two* daughters?" This is because, with the food and other goods needed to rear one son, you could have reared two daughters. The evolutionarily stable sex ratio, measured in numbers of bodies, would then be two females to every male. But *measured in amounts of parental expenditure* (as opposed to numbers of individuals), the evolutionarily stable sex ratio is still 50:50. Fisher's theory amounts to a balancing of the expenditures on the two sexes. This often, as it happens, turns out to be the same as balancing the numbers of the two sexes.

Even in seals, as I said, it looks as though the amount of parental expenditure on sons is not noticeably different from the amount spent on daughters. The massive inequality in weight seems to come about after the end of parental expenditure. So the decision facing a parent is still "Should I have a son or a daughter?" Even though the total cost of a son's growth to adulthood may be much more than the total cost of a daughter's growth, if the additional cost is not borne by the decision maker (the parent) that's all that counts in Fisher's theory.

Fisher's rule about balancing the expenditure still holds in those cases where one sex suffers a higher rate of mortality than the other. Suppose, for instance, that male babies are more likely to die than female babies. If the sex ratio at conception is exactly 50:50, the males reaching adulthood will be outnumbered by the females. They are therefore the minority sex, and we'd naively expect natural selection to favor parents that specialize in sons. Fisher would expect this too, but only up to a point—and a precisely limited point, at that. He would not expect parents to conceive such a surplus of sons that the greater infant mortality is exactly compensated, leading to equality in the breeding population. No, the sex ratio at conception should be somewhat male-biased, but only up to the point where the total expenditure on sons is expected to equal the total expenditure on daughters.

Once again, the easiest way to think about it is to put yourself in the position of the decision-making parent and ask the question "Should I have a daughter, who will probably survive, or a son, who may die in infancy?" The decision to make grandchildren via sons entails a probability that you'll have to spend more resources on some extra sons to replace those that are going to die. You can think of each surviving son as carrying the ghosts of his dead brothers on his back. He carries them on his back in the sense that the decision to go the son route to grandchildren lets the parent in for some additional wasted expenditure—expenditure that will be squandered on dead infant males. Fisher's fundamental rule still holds good. The total amount of goods and energy invested in sons (including feeding infant sons up to the point where they died) will equal the total amount invested in daughters.

What if, instead of higher male infant mortality, there is higher male mortality after the end of parental expenditure? In fact this will often be so, because adult males often fight and injure each other. This circumstance, too, will lead to a surplus of females in the breeding population. On the face of it, therefore, it would seem to favor parents who specialize in sons, thereby taking advantage of the rarity of males in the breeding population. Think a little harder, however, and you realize that the reasoning is fallacious. The decision facing a parent is the following: "Should I have a son, who will likely be killed in battle after I've finished rearing him but who, if he survives, will give me extra specially many grandchildren? Or shall I have a daughter, who is fairly certain to give me an average number of grandchildren?" The number of grandchildren you can expect through a son is still the same as the average number you can expect through a daughter. And the cost of making a son is still the cost of feeding and protecting him up to the moment when he leaves the nest. The fact that he is likely to get killed soon after he leaves the nest does not change the calculation.

In all this reasoning, Fisher assumed that the "decision maker" is the parent. The calculation changes if it is somebody else. Suppose, for instance, that an individual could influence its own sex. Once again, I don't mean influence by conscious intention. I am hypothesizing genes that switch an individual's development into the female or the male pathway, conditional upon cues from the environment. Following our usual convention, for brevity I shall use the language of deliberate choice by an individual—in this case, deliberate choice of its own sex. If harem-based animals like elephant seals were granted this power of flexible choice, the effect would be dramatic. Individuals would aspire to be harem-holding males, but if they failed at acquiring a harem they would much prefer to be females than bachelor males. The sex ratio in the population would become strongly female-biased. Elephant seals unfortunately can't reconsider the sex they were given at conception, but some fish can. Males of the blue-headed wrasse are large and bright-colored, and they hold harems of dull-colored females. Some females are larger than others, and they form a dominance hierarchy. If a male dies his place is quickly taken by the largest female, who soon turns into a bright-colored male. These fish get the best of both worlds. Instead of wasting their lives as bachelor males waiting for the death of a dominant, harem-holding male, they spend their waiting time as productive females. The blue-headed wrasse sex-ratio system is a rare one, in which God's Utility Function coincides with something that a social economist might regard as prudent.

So, we've considered both the parent and the self as decision maker. Who else might the decision maker be? In the social insects the investment decisions are made, in large part, by sterile workers, who will normally be elder sisters (and also brothers, in the case of termites) of the young being reared. Among the more familiar social insects are honeybees. Beekeepers among my readers may already have recognized that the sex ratio in the hive doesn't seem, on the face of it, to conform to Fisher's expectations. The first thing to note is that workers should not be counted as females. They are technically females, but they don't reproduce, so the sex ratio being regulated according to Fisher's theory is the ratio of drones (males) to new queens being churned out by the hive. In the case of bees and ants, there are special technical reasons, which I have discussed in *The Selfish Gene* and won't rehearse here, for expecting the sex ratio to be 3:1 in favor of females. Far from this, as any beekeeper knows, the actual sex ratio is heavily male-biased. A flourishing hive may produce half a dozen new queens in a season but hundreds or even thousands of drones.

What is going on here? As so often in modern evolutionary theory, we owe the answer to W.D. Hamilton, now at Oxford University. It is revealing and epitomizes the whole Fisher-inspired theory of sex ratios. The key to the riddle of bee sex ratios lies in the remarkable phenomenon of swarming. A beehive is, in many ways, like a single individual. It grows to maturity, it reproduces, and eventually it dies. The reproductive product of a beehive is a swarm. At the height of summer, when a hive has been really prospering, it throws off a daughter colony—a swarm. Producing swarms is the equivalent of reproduction, for the hive. If the hive is a factory, swarms are the end product, carrying with them the precious genes of the colony. A swarm comprises one queen and several thousand workers. They all leave the parent hive in a body and gather as a dense cluster, hanging from a bough or a rock. This will be their temporary encampment while they prospect for a new permanent home. Within a few days, they find a cave or a hollow tree (or, more usually nowadays, they are captured by a beekeeper, perhaps the original one, and housed in a new hive).

It is the business of a prosperous hive to throw off daughter swarms. The first step in doing this is to make a new queen. Usually half a dozen or so new queens are made, only one of whom is destined to live. The first one to hatch stings all the others to death. (Presumably the surplus queens are there only for insurance purposes.) Queens are genetically interchangeable with workers, but they are reared in special queen cells that hang below the comb, and they are fed on a specially rich, queen-nourishing diet. This diet includes royal jelly, the substance to which the novelist Dame Barbara Cartland romantically attributes her long life and queenly deportment. Worker bees are reared in smaller cells, the same cells that are later used to store honey. Drones are genetically different. They come from unfertilized eggs. Remarkably, it is up to the queen whether an egg turns into a drone or into a female (queen/worker). A queen bee mates only during a single mating flight, at the beginning of her adult life, and she stores the sperm for the rest of her life, inside her body. As each egg passes down her egg tube, she may or may not release a small package of sperm from her store, to fertilize it. The queen, therefore, is in control of the sex ratio among eggs. Subsequently, however, the workers seem to have all the power, because they control the food supply for the larvae. They could, for instance, starve male larvae if the queen laid too many (from their point of view) male eggs. In any case the workers have control over whether a female egg turns into a worker or a queen, since this depends solely on rearing conditions, especially diet.

Now let's return to our sex-ratio problem and look at the decision facing the workers. As we have seen, unlike the queen, they are not choosing whether to produce sons or daughters but whether to produce brothers (drones) or sisters (young queens). And now we are back to our riddle. For the actual sex ratio seems to be massively male-biased, which doesn't seem to make sense from Fisher's point of view. Let's look harder at the decision facing the workers. I said that it was a choice between brothers and sisters. But wait a moment. The decision to rear a brother is, indeed, just that: it commits the hive to whatever food and other resources it takes to rear one drone bee. But the decision to rear a new queen commits the hive to far more than than just the resources needed to nourish one queen's body. The decision to rear a new queen is tantamount to a commitment to lay down a swarm. The true cost of a new queen only negligibly includes the small amount of royal jelly and other food that she will eat. It mostly consists of the cost of making all the thousands of workers who are going to be lost to the hive when the swarm departs.

This is almost certainly the true explanation for the apparently anomalous male bias in the sex ratio. It turns out to be an extreme example of what I was talking about earlier. Fisher's rule states that the quantity of expenditure on males and females must be equal, not the census count of male and female individuals. The expenditure on a new queen entails massive expenditure on workers who would not otherwise have been lost to the hive. It is like our hypothetical seal population, in which one sex costs twice as much as the other to rear, with the result that that sex is half as numerous. In the case of bees a queen costs hundreds or even thousands of times as much as a drone, because she carries on her back the cost of all the extra workers needed for the swarm. Therefore queens are hundreds of times less numerous than drones. There is an additional sting to this curious tale: when a swarm leaves, it mysteriously contains the *old* queen, not the new one. Nevertheless, the economics are the same. The decision to make a new queen still entails the outlay of the swarm needed to escort the old queen to her new home.

To round off our treatment of sex ratios, we return to the

riddle of the harems with which we began: that profligate arrangement whereby a large herd of bachelor males consumes nearly half (or even more than half) the population's food resources but never reproduces nor does anything else useful. Obviously the economic welfare of the population is not being maximized here. What is going on? Once again, put yourself in the position of the decision maker—say, a mother trying to "decide" whether to have a son or a daughter in order to maximize the number of her grandchildren. Her decision is, at naive first sight, an unequal one: "Should I have a son, who will probably end up a bachelor and give me no grandchildren at all, or a daughter, who will probably end up in a harem and will give me a respectable number of grandchildren?" The proper reply to this would-be parent is "But if you have a son, he *may* end up with a harem, in which case he'll give you far more grandchildren than you could ever hope to get via a daughter." Suppose, for simplicity, that all the females reproduce at the average rate, and that nine out of ten males never reproduce, while one male in ten monopolizes the females. If you have a daughter, you can count on an average number of grandchildren. If you have a son, you have a 90 percent chance of no grandchildren but a 10 percent chance of having ten times the average number of grandchildren. The average number of grandchildren you can expect through your sons is the same as the average number you can expect through your daughters. Natural selection still favors a 50:50 sex ratio, even though species-level economic reason cries out for a surplus of females. Fisher's rule still holds.

I expressed all these reasonings in terms of "decisions" of individual animals but, to repeat, this is just shorthand. What is really going on is that genes "for" maximizing grandchildren become more numerous in the gene pool. The world becomes full of genes that have successfully come down the ages. How should a gene be successful in coming down the ages other than by influencing the decisions of individuals so as to maximize their numbers of descendants? Fisher's sex-ratio theory tells us how this maximizing should be done, and it is very different from maximizing the economic welfare of the species or population. There is a utility function here, but it is far from the utility function that would spring to our human economic minds.

The wastefulness of the harem economy can be summarized as follows: Males, instead of devoting themselves to useful work, squander their energy and strength in futile struggles against one another. This is true, even if we define "useful" in an apparently Darwinian way, as concerned with rearing children. If males diverted into useful channels the energy that they waste competing with each other, the species as a whole would rear more children for less effort and less food consumed.

A work-study expert would stare aghast at the world of the elephant seal. An approximate parallel would be the following. A workshop needs no more than ten men to run it, since there are just ten lathes in the workshop. Instead of simply employing ten men, the management decides to employ a hundred men. Every day, all hundred men turn up and collect their wages. Then they spend the day fighting for possession of the ten lathes. Some items get made on the lathes, but no more than would have been achieved by ten men, and probably fewer, because the hundred men are so busy fighting that the lathes are not being used efficiently.

The work-study expert would be in no doubt. Ninety percent of the men are redundant, and they should be officially declared so and dismissed.

It isn't just in physical combat that male animals waste their efforts—"waste" being defined, once again, from the point of view of the human economist or work-study expert. In many species there's a beauty contest too. This brings us to another utility function that we humans can appreciate even though it doesn't make straightforward economic sense: aesthetic beauty. On the face of it, it might look as though God's Utility Function is sometimes drawn up along the lines of the (now thankfully unfashionable) Miss World contest, but with males parading the runway. This is seen most clearly in the so-called leks of birds such as grouse and ruffs. A "lek" is a patch of ground traditionally used by male birds for parading in front of females. Females visit the lek and watch the swaggering displays of a number of males before singling one out and copulating with him. The males of lekking species often have bizarre ornamentation, which they show off with equally remarkable bowing or bobbing movements and strange noises. The word "bizarre" is, of course, a subjective value judgment; presumably lekking male sage grouse, with their puffed-up dances accompanied by cork-popping noises, don't seem bizarre to the females of their own species, and this is all that matters. In some cases the female birds' idea of beauty happens to coincide with ours, and the result is a peacock or a bird of paradise.

Nightingale songs, pheasant tails, firefly flashes and the rainbow scales of tropical reef fish are all maximizing aesthetic beauty, but it is not—or is only incidentally—beauty for human delectation. If we enjoy the spectacle it is a bonus, a by-product. Genes that make males attractive to females automatically find themselves passed down the digital river to the future. There is only one utility function that makes sense of these beauties; it is the same one that explains elephant-seal sex ratios, cheetahs and antelopes running superficially futile races against each other, cuckoos and lice, eyes and ears and windpipes, sterile worker ants and superfertile queens. The great universal Utility Function, the quantity that is being diligently maximized in every cranny of the living world is, in every case, the survival of the DNA responsible for the feature you are trying to explain.

Peacocks are burdened with finery so heavy and cumbersome that it would gravely hamper their efforts to do useful work, even if they felt inclined to do useful work—which, on the whole, they don't. Male songbirds use dangerous amounts of time and energy singing. This certainly imperils them, not only because it attracts predators but because it drains energy and uses time that could be spent replenishing that energy. A student of wren biology claimed that one of his wild males sang itself literally to death. Any utility function that had the long-term welfare of the species at heart, even the long-term survival of this particular individual male, would cut down on the amount of singing, the amount of displaying, the amount of fighting among males. Yet, because what is really being maximized is DNA survival, nothing can stop the spread of DNA that has no beneficial effect other than making males beautiful to females. Beauty is not an absolute virtue in itself. But inevitably, if some genes do confer on males whatever qualities the females of the species happen to find desirable, those genes, willy-nilly, will survive.

Why are forest trees so tall? Simply to overtop rival trees. A "sensible" utility function would see to it that they were all short. They would get exactly the same amount of sunlight, with far less expenditure on thick trunks and massive supporting buttresses. But if they were all short, natural selection couldn't help favoring a variant individual that grew a little taller. The ante having been upped, others would have to follow suit. Nothing can stop the whole game escalating until all trees are ludicrously and wastefully tall. It is ludicrous and wasteful only from the point of view of a rational economic planner thinking in terms of maximizing efficiency. But it all makes sense once you understand the true utility function—genes are maximizing their own survival. Homely analogies abound. At a cocktail party, you shout yourself hoarse. The reason is that everybody else is shouting at top volume. If only the guests could come to an agreement to whisper, they'd hear one another exactly as well with less voice strain and less expenditure of energy. But agreements like that don't work unless they are policed. Somebody always spoils it by selfishly talking a bit louder, and, one by one, everybody has to follow suit. A stable equilibrium is reached only when everybody is shouting as loudly as physically possible, and this is much louder than required from a "rational" point of view. Time and again, cooperative restraint is thwarted by its own internal instability. God's Utility Function seldom turns out to be the greatest good for the greatest number. God's Utility Function betrays its origins in an uncoordinated scramble for selfish gain.

Humans have a rather endearing tendency to assume that welfare means group welfare, that "good" means the good of society, the future well-being of the species or even of the ecosystem. God's Utility Function, as derived from a contemplation of the nuts and bolts of natural selection, turns out to be sadly at odds with such utopian visions. To be sure, there are occasions when genes may maximize their selfish welfare at their level, by programming unselfish cooperation, or even self-sacrifice, by the organism at its level. But group welfare is always a fortuitous consequence, not a primary drive. This is the meaning of "the selfish gene."

Let us look at another aspect of God's Utility Function, beginning with an analogy. The Darwinian psychologist Nicholas Humphrey made up an illuminating fact about Henry Ford. "It is said" that Ford, the patron saint of manufacturing efficiency, once

> commissioned a survey of the car scrapyards of America to find out if there were parts of the Model T Ford which never failed. His inspectors came back with reports of almost every kind of breakdown: axles, brakes, pistons—all were liable to go wrong. But they drew attention to one notable exception, the *kingpins* of the scrapped cars invariably had years of life left in them. With ruthless logic Ford concluded that the kingpins on the Model T were too good for their job and ordered that in future they should be made to an inferior specification.

You may, like me, be a little vague about what kingpins are, but it doesn't matter. They are something that a motor car needs, and Ford's alleged ruthlessness was, indeed, entirely logical. The alternative would have been to improve all the other bits of the car to bring them up to the standard of the kingpins. But then it wouldn't have been a Model T he was manufacturing but a Rolls Royce, and that wasn't the object of the exercise. A Rolls Royce is a respectable car to manufacture and so is a Model T, but for a different price. The trick is to make sure that either the whole car is built to Rolls Royce specifications or the whole car is built to Model T specifications. If you make a hybrid car, with some components of Model T quality and some components of Rolls Royce quality, you are getting the worst of both worlds, for the car will be thrown away when the weakest of its components wears out, and the money spent on high-quality components that never get time to wear out is simply wasted.

Ford's lesson applies even more strongly to living bodies than to cars, because the components of a car can, within limits, be replaced by spares. Monkeys and gibbons make their living in the treetops and there is always a risk of falling and breaking bones. Suppose we commissioned a survey of monkey corpses to count the frequency of breakage in each major bone of the body. Suppose it turned out that every bone breaks at some time or another, with one exception: the fibula (the bone that runs parallel to the shinbone) has never ever been observed to break in any monkey. Henry Ford's unhesitating prescription would be to redesign the fibula to an inferior specification, and this is exactly what natural selection would do too. Mutant individuals with an inferior fibula—mutant individuals whose growth rules call for diverting precious calcium away from the fibula—could use the material saved to thicken other bones in the body and so achieve the ideal of making every bone equally likely to break. Or the mutant individuals could use the calcium saved to make more milk and so rear more young. Bone can safely be shaved off the fibula, at least up to the point where it becomes as likely to break as the next most durable bone. The alternative—the "Rolls Royce" solution of bringing all the other components up to the standard of the fibula—is harder to achieve.

The calculation isn't quite as simple as this, because some bones are more important than others. I guess it is easier for a spider monkey to survive with a fractured heelbone than with a fractured armbone, so we should not literally expect natural selection to make all bones equally likely to break. But the main lesson we take away from the legend of Henry Ford is undoubtedly correct. It is possible for a component of an animal to be too good, and we should expect natural selection to favor a lessening of quality up to, but not beyond, a point of balance with the quality of the other components of the body. More precisely, natural selection will favor a leveling out of quality in both the downward and upward directions, until a proper balance is struck over all parts of the body.

It is especially easy to appreciate this balance when it is struck between two rather separate aspects of life: peacock survival versus beauty in the eyes of peahens, for instance. Darwinian theory tells us that all survival is just a means to the end of gene propagation, but this does not stop us partitioning the body into those components, like legs, that are primarily concerned with individual survival and those, like penises, that are concerned with reproduction. Or those, like antlers, that are devoted to competing with rival individuals versus those, like legs and penises, whose importance does not depend upon the existence of rival individuals. Many insects impose a rigid separation between radically different stages in their life history.

Caterpillars are devoted to gathering food and growing. Butterflies are like the flowers they visit, devoted to reproducing. They do not grow, and they suck nectar only to burn it immediately as aviation fuel. When a butterfly reproduces successfully, it spreads the genes not just for being an efficient flying and mating butterfly but for being the efficient feeding caterpillar that it was, as well. Mayflies feed and grow as underwater nymphs for up to three years. They then emerge as flying adults that live only a matter of hours. Many of them are eaten by fish, but even if they were not they would soon die anyway, because they cannot feed and they do not even possess guts (Henry Ford would have loved them). Their job is to fly until they find a mate. Then, having passed on their genes—including the genes for being an efficient nymph capable of feeding underwater for three years—they die. A mayfly is like a tree that takes years to grow, then flowers for a single glorious day and dies. The adult mayfly is the flower that briefly blooms at the end of life and the beginning of new life.

A young salmon migrates down the stream of its birth and spends the bulk of its life feeding and growing in the sea. When it reaches maturity it again seeks out, probably by smell, the mouth of its native stream. In an epic and much-celebrated journey the salmon swims upstream, leaping falls and rapids, home to the headwaters from which it sprang a lifetime ago. There it spawns and the cycle renews. At this point there is typically a difference between Atlantic and Pacific salmon. The Atlantic salmon, having spawned, may return to the sea with some chance of repeating the cycle a second time. Pacific salmon die, spent, within days of spawning.

A typical Pacific salmon is like a mayfly but without the anatomically clear-cut separation between nymph and adult phases in the life history. The effort of swimming upstream is so great that it cannot pay to do it twice. Therefore natural selection favors individuals that put every ounce of their resources into one "big bang" reproductive effort. Any resources left after breeding would be wasted—the equivalent of Henry Ford's overdesigned kingpins. The Pacific salmon have evolved toward whittling down their postreproductive survival until it approaches zero, the resources saved being diverted into eggs or milt. The Atlantic salmon were drawn toward the other route. Perhaps because the rivers they have to mount tend to be shorter and spring from less formidable hills, individuals that keep some resources back for a second reproductive cycle can sometimes do well by it. The price these Atlantic salmon pay is that they cannot commit so much to their spawn. There is a trade-off between longevity and reproduction, and different kinds of salmon have opted for different equilibria. The special feature of the salmon life cycle is that the grueling odyssey of their migration imposes a discontinuity. There is no easy continuum between one breeding season and two. Commitment to a second breeding season drastically cuts into efficiency in the first. Pacific salmon have evolved toward an unequivocal commitment to the first breeding season, with the result that a typical individual unequivocally dies immediately after its single titanic spawning effort.

The same kind of trade-off marks every life, but it is usually less dramatic. Our own death is probably programmed in something like the same sense as that of the salmon but in a less downright and clear-cut fashion. Doubtless a eugenicist could breed a race of superlatively long-lived humans. You would choose for breeding those individuals who put most of their resources into their own bodies at the expense of their children: individuals, for example, whose bones are massively reinforced and hard to break but who have little calcium left over to make milk. It is easy enough to live a bit longer, if you are cosseted at the expense of the next generation. The eugenicist could do the cosseting and exploit the trade-offs in the desired direction of longevity. Nature will not cosset in this way, because genes for scrimping the next generation will not penetrate the future.

Nature's Utility Function never values longevity for its own sake but only for the sake of future reproduction. Any animal that, like us but unlike a Pacific salmon, breeds more than once faces trade-offs between the current child (or litter) and future children. A rabbit that devoted all her energy and resources to her first litter would probably have a superior first litter. But she would have no resources left to carry her on to a second litter. Genes for keeping something in reserve will tend to spread through the rabbit population, carried in the bodies of second- and third-litter babies. It is genes of this kind that so conspicuously did not spread through the population of Pacific salmon, because the practical discontinuity between one breeding season and two is so formidable.

As we grow older our chances of dying within the next year, after initially decreasing and then plateauing for a while, settle down to a long climb. What is happening in this long increase in mortality? It is basically the same principle as for the Pacific salmon, but spread out over an extended period instead of being concentrated in a brief precipitous orgy of death after the orgy of spawning. The principle of how senescence evolved was originally worked out by the Nobel laureate and medical scientist Sir Peter Medawar in the early 1950s, with various modifications to the basic idea added by the distinguished Darwinians G.C. Williams and W.D. Hamilton.

The essential argument is as follows: First, as we saw in chapter 1, any genetic effect will normally be switched on at a particular time during the life of the organism. Many genes are switched on in the early embryo, but others—like the gene for Huntington's chorea, the disease that tragically killed the folk poet and singer Woody Guthrie—are not switched on until middle age. Second, the details of a genetic effect, including the time at which it is switched on, may be modified by other genes. A man possessing the Huntington's chorea gene can expect to die from the disease, but whether it kills him when he is forty or when he is fifty-five (as Woody Guthrie was) may be influenced by other genes. It follows that by selection of "modifier" genes the time of action of a particular gene can either be postponed or brought forward in evolutionary time.

A gene like the Huntington's chorea gene, which switches on between the ages of thirty-five and fifty-five, has plenty of opportunity to be passed on to the next generation before it kills its possessor. If, however, it were switched on at the age of twenty, it would be passed on only by people who reproduce rather young, and therefore it would be strongly selected against. If it were switched on at the age of ten, it would essentially never be passed on. Natural selection would favor any modifier genes that had the effect of postponing the age of switching on of the Huntington's chorea gene. According to the Medawar/Williams theory, this would be exactly why it normally does not switch on until middle age. Once upon a time it may well have been an early maturing gene, but natural selection has favored a postponing of its lethal effect until middle age. No doubt there is still slight selection pressure to push it on into old age, but this pressure is weak because so few victims die before reproducing and passing the gene on.

The Huntington's chorea gene is a particularly clear example of a lethal gene. There are lots of genes that are not in themselves lethal but nevertheless have effects that increase the probability of dying from some other cause and are called sublethal. Once again, their time of switching on may be influenced by modifier genes and therefore postponed or accelerated by natural selection. Medawar realized that the debilities of old age might represent an accumulation of lethal and sublethal genetic effects that had been pushed later and later in the life cycle and allowed to slip through the reproductive net into future generations simply because they were late-acting.

The twist that G. C. Williams, the doyen of modern American Darwinists, gave to the story in 1957 is an important one. It gets back to our point about economic trade-offs. To understand it, we need to throw in a couple of additional background facts. A gene usually has more than one effect, often on parts of the body that are superficially quite distinct. Not only is this "pleiotropy" a fact, it is also very much to be expected, given that genes exert their effects on embryonic development and embryonic development is a complicated process. So, any new mutation is likely to have not just one effect but several. Though one of its effects may be beneficial, it is unlikely that more than one will be. This is simply because most mutational effects are bad. In addition to being a fact, this is to be expected in principle: if you start with a complicated working mechanism—like a radio, say—there are many more ways of making it worse than of making it better.

Whenever natural selection favors a gene because of its beneficial effect in youth—say, on sexual attractiveness in a young male—there is likely to be a downside: some particular disease in middle or old age, for example. Theoretically, the age effects could be the other way around but, following the Medawar logic, natural selection is hardly going to favor disease in the young because of a beneficial effect of the same gene in old age. Moreover, we can invoke the point about modifier genes again. Each of the several effects of a gene, its good and its bad effects, could have their switch-on times altered in subsequent evolution. According to the Medawar principle, the good effects would tend to be moved earlier in life, while the bad effects would tend to be postponed until later. Moreover, there will in some cases be a direct trade-off between early and late effects. This was implied in our discussion of salmon. If an animal has a finite quantity of resources to spend on, say, becoming physically strong and able to leap out of danger, any predilection to spend those resources early will be favored over a preference to spend them late. Late spenders are more likely to be already dead from other causes before they have a chance to spend their resources. To put the general Medawar point in a sort of back-to-front version of the language we introduced in chapter 1, everybody is descended from an unbroken line of ancestors all of whom were at some time in their lives young but many of whom were never old. So we inherit whatever it takes to be young, but not necessarily whatever it takes to be old. We tend to inherit genes for dying a long time after we're born, but not for dying a short time after we're born.

To return to this chapter's pessimistic beginning, when the utility function—that which is being maximized—is DNA survival, this is not a recipe for happiness. So long as DNA is passed on, it does not matter who or what gets hurt in the process. It is better for the genes of Darwin's ichneumon wasp that the caterpillar should be alive, and therefore fresh, when it is eaten, no matter what the cost in suffering. Genes don't care about suffering, because they don't care about anything.

If Nature were kind, she would at least make the minor concession of anesthetizing caterpillars before they are eaten alive from within. But Nature is neither kind nor unkind. She is neither against suffering nor for it. Nature is not interested one way or the other in suffering, unless it affects the survival of DNA. It is easy to imagine a gene that, say, tranquilizes gazelles when they are about to suffer a killing bite. Would such a gene be favored by natural selection? Not unless the act of tranquilizing a gazelle improved that gene's chances of being propagated into future generations. It is hard to see why this should be so, and we may therefore guess that gazelles suffer horrible pain and fear when they are pursued to the death—as most of them eventually are. The total amount of suffering per year in the natural world is beyond all decent contemplation. During the minute it takes me to compose this sentence, thousands of animals are being eaten alive; others are running for their lives, whimpering with fear; others are being slowly devoured from within by rasping parasites; thousands of all kinds are dying of starvation, thirst and disease. It must be so. If there is ever a time of plenty, this very fact will automatically lead to an increase in population until the natural state of starvation and misery is restored.

Theologians worry away at the "problem of evil" and a related "problem of suffering." On the day I originally wrote this paragraph, the British newspapers all carried a terrible story about a bus full of children from a Roman Catholic school that crashed for no obvious reason, with wholesale loss of life. Not for the first time, clerics were in paroxysms over the theological question that a writer on a London newspaper (*The Sunday Telegraph*) framed this way: "How can you believe in a loving, all-powerful God who allows such a tragedy?" The article went on to quote one priest's reply: "The simple answer is that we do not know why there should be a God who lets these awful things happen. But the horror of the crash, to a Christian, confirms the fact that we live in a world of real values: positive and negative. If the universe was just electrons, there would be no problem of evil or suffering."

On the contrary, if the universe were just electrons and selfish genes, meaningless tragedies like the crashing of this bus are exactly what we should expect, along with equally meaningless *good* fortune. Such a universe would be neither evil nor good in intention. It would manifest no intentions of any kind. In a universe of blind physical forces and genetic replication, some people are going to get hurt, other people are going to get lucky, and you won't find any rhyme or reason in it, nor any justice. The universe we observe has precisely the properties we should expect if there is, at bottom, no design, no purpose, no evil and no good, nothing but blind, pitiless indifference. As that unhappy poet A. E. Housman put it:

> For Nature, heartless, witless Nature
> Will neither know nor care.

DNA neither knows nor cares. DNA just is. And we dance to its music.

CHAPTER 5

THE REPLICATION BOMB

Most stars—and our sun is typical—burn in a stable manner for thousands of millions of years. Very rarely, somewhere in the galaxy a star suddenly flares without obvious warning into a supernova. Within a period of a few weeks, it increases in brightness by a factor of many billions and then dies away to a dark remnant of its former self. During its few high days as a supernova, a star may radiate more energy than in all its previous hundred million years as an ordinary star. If our own sun were to "go supernova," the entire solar system would be vaporized on the instant. Fortunately this is very unlikely. In our galaxy of a hundred billion stars, only three supernovas have been recorded by astronomers: in 1054, in 1572, and in 1604. The Crab Nebula is the remains of the event of 1054, recorded by Chinese astronomers. (When I say the event "of 1054" I mean, of course, the event of which news reached Earth in 1054. The event itself took place six thousand years earlier. The wavefront of light from it hit us in 1054.) Since 1604, the only supernovas that have been seen have been in other galaxies.

There is another type of explosion a star can sustain. Instead of "going supernova" it "goes information." The explosion begins more slowly than a supernova and takes incomparably longer to build up. We can call it an information bomb or, for reasons that will become apparent, a replication bomb. For the first few billion years of its build-up, you could detect a replication bomb only if you were in the immediate vicinity. Eventually, subtle manifestations of the explosion begin to leak away into more distant regions of space and it becomes, at least potentially, detectable from a long way away. We do not know how this kind of explosion ends. Presumably it eventually fades away like a supernova, but we do not know how far it typically builds up first. Perhaps to a violent and self-destructive catastrophe. Perhaps to a more gentle and repeated emission of objects, moving, in a guided rather than a simple ballistic trajectory, away from the star into distant reaches of space, where it may infect other star systems with the same tendency to explode.

The reason we know so little about replication bombs in the universe is that we have seen only one example, and one example of any phenomenon is not enough to base generalizations on. Our one case history is still in progress. It has been under way for between three billion and four billion years, and it has only just reached the threshold of spilling away from the immediate vicinity of the star. The star concerned is Sol, a yellow dwarf star lying toward the edge of our galaxy, in one of the spiral arms. We call it the sun. The explosion actually originated on one of the satellites in close orbit around the sun, but the energy to drive the explosion all comes from the sun. The satellite is, of course, Earth, and the four-billion-year-old explosion, or replication bomb, is called life. We humans are an extremely important manifestation of the replication bomb, because it is through us—through our brains, our symbolic culture and our technology—that the explosion may proceed to the next stage and reverberate through deep space.

As I have said, our replication bomb is, to date, the only one we know of in the universe, but this does not necessarily mean that events of this kind are rarer than supernovas. Admittedly, supernovas have been detected three times as frequently in our galaxy, but then supernovas, because of the immense quantities of energy released, are much easier to see from a long distance. Until a few decades ago, when man-made radio waves started to radiate outward from the planet, our own life explosion would have gone undetected by observers even on quite close planets. Probably the only conspicuous manifestation of our life explosion until recent times would have been the Great Barrier Reef.

A supernova is a gigantic and sudden explosion. The triggering event of any explosion is that some quantity is tipped over a critical value, after which things escalate out of control to produce a result far larger than the original triggering event. The triggering event of a replication bomb is the spontaneous arising of self-replicating yet variable entities. The reason self-replication is a potentially explosive phenomenon is the same as for any explosion: exponential growth—the more you have, the more you get. Once you have a self-replicating object, you will soon have two. Then each of the two makes a copy of itself and then you have four. Then eight, then sixteen, thirty-two, sixty-four. . . . After a mere thirty generations of this duplication, you will have more than a billion of the duplicating objects. After fifty generations, there will be a thousand million million of them. After two hundred generations, there will be a million million million million million million million million million million. In theory. In practice it could never come to pass, because this is a larger number than there are atoms in the universe. The explosive process of self-copying has got to be limited long before it reaches two hundred generations of unfettereddoubling.

We have no direct evidence of the replication event that initiated the proceedings on this planet. We can only infer that it must have happened because of the gathering explosion of which we are a part. We do not know exactly what the original critical event, the initiation of self-replication, looked like, but we can infer what kind of an event it must have been. It began as a chemical event.

Chemistry is a drama that goes on in all stars and on all planets. The players in chemistry are atoms and molecules. Even the rarest of atoms are extremely numerous by the standards of counting to which we are accustomed. Isaac Asimov calculated that the number of atoms of the rare element astatine-215 in the whole of North and South America to a depth of ten miles is "only a trillion." The fundamental units of chemistry are forever changing partners to produce a shifting but always very large population of larger units—molecules. However numerous they are, molecules of a given type—unlike, say, animals of a given species or Stradivarius violins—are always identical. The atomic dance routines of chemistry lead to some molecules becoming more populous in the world while others become scarcer. A biologist is naturally tempted to describe the molecules that become more numerous in the population as "successful." But it is not helpful to succumb to this temptation. Success, in the illuminating sense of the word, is a property that arises only at a later stage in our story.

What, then, was this momentous critical event that began the life explosion? I have said that it was the arising of self-duplicating entities, but equivalently we could call it the origination of the phenomenon of heredity—a process we can label "like begets like." This is not something molecules ordinarily exhibit. Water molecules, though they swarm in gigantic populations, show nothing approaching true heredity. On the face of it, you might think they do. The population of water molecules (H_2O) increases when hydrogen (H) burns with oxygen (O). The population of water molecules decreases when water is split, by electrolysis, into bubbles of hydrogen and oxygen. But although there is a kind of population dynamics of water molecules, there is no heredity. The minimal condition for true heredity would be the existence of at least two distinct kinds of H_2O molecule, both of which give rise to ("spawn") copies of their own kind.

Molecules sometimes come in two mirror varieties. There are two kinds of glucose molecule, which contain identical atoms Tinkertoyed together in an identical way except that the molecules are mirror images. The same is true of other sugar molecules, and lots of other molecules besides, including the all-important amino acids. Perhaps here is an opportunity for "like begets like"—for chemical heredity. Could right-handed molecules spawn right-handed daughter molecules and left-handers spawn southpaw daughter molecules? First, some background information on mirror-image molecules. The phenomenon was first discovered by the great nineteenth-century French scientist Louis Pasteur, who was looking at crystals of tartrate, which is a salt of tartaric acid, an important substance in wine. A crystal is a solid edifice, big enough to be seen with the naked eye and, in some cases, worn around the neck. It is formed when atoms or molecules, all of the same type, pile on top of one another to form a solid. They don't pile up hugger-mugger but in an orderly geometric array, like guardsmen of identical size and immaculate drill. The molecules that are already part of the crystal constitute a template for the addition of new molecules, which come out of a watery solution and fit it exactly, so the whole crystal grows as a precise, geometric lattice. This is why salt crystals have square facets and diamond crystals are tetrahedral (diamond-shaped). When any shape acts as a template for building another shape like itself, we have an inkling of the possibility of self-replication.

Now, back to Pasteur's tartrate crystals. Pasteur noticed that when he left a solution of tartrate in water, *two* different kinds of crystal emerged, identical except that they were mirror images of each other. He laboriously sorted the two kinds of crystal into two separate heaps. When he separately redissolved them, he obtained two different solutions, two kinds of tartrate in solution. Although the two solutions were similar in most respects, Pasteur found that they rotated polarized light in opposite directions. It is this that gives the two kinds of molecule their conventional names of left- and right-handed, since they rotate polarized light anticlockwise and clockwise, respectively. As you would guess, when the two solutions were allowed to crystallize out once more, each produced pure crystals that mirrored the pure crystals of the other.

Mirror-image molecules really are distinct in that, as

with left and right shoes, no matter how hard you try, you can't rotate them so that one can be used as a substitute for the other. Pasteur's original solution was a mixed population of two kinds of molecules, and the two kinds each insisted on lining up with their own kind when crystallizing out. The existence of two (or more) distinct varieties of an entity is a necessary condition for there to be true heredity, but it is not sufficient. For there to be true heredity among the crystals, left- and right-handed crystals would have to split in half when they reached a critical size and each half serve as a template for growth to full size again. Under these conditions we really would have a growing population of two rival kinds of crystals. We truly might speak of "success" in the population, because—since both types are competing for the same constituent atoms—one type might become more numerous at the expense of the other, by virtue of being "good" at making copies of itself. Unfortunately, the vast majority of known molecules do not have this singular property of heredity.

I say "unfortunately" because chemists, trying for medical purposes to make molecules that are all, say, left-handed, would dearly like to be able to "breed" them. But insofar as molecules act as templates for the formation of other molecules, they normally do so for their mirror image, not for their like-handed form. This makes things difficult, because if you start with a left-handed form you end up with an equal mixture of left- and right-handed molecules. Chemists involved in this field are trying to trick molecules into "breeding" daughter molecules of the same handedness. It is a very difficult trick to pull off.

In effect, though it probably didn't involve handedness, a version of this trick was pulled off naturally and spontaneously four thousand million years ago, when the world was new and the explosion that turned into life and information began. But something more than simple heredity was needed before the explosion could properly get under way. Even if a molecule does show true heredity among left-handed and right-handed forms, any competition between them would not have very interesting consequences, because there are only two kinds. Once the left-handers, say, had won the competition, that would be the end of the matter. There would be no more progress.

Larger molecules can exhibit handedness at different parts of the molecule. The antibiotic monensin, for instance, has seventeen centers of asymmetry. At every one of these seventeen centers, there is a left-handed and a right-handed form. Two multiplied by itself 17 times is 131,072, and there are therefore 131,072 distinct forms of the molecule. If these 131,072 possessed the property of true heredity, with each one begetting only its own kind, there could be quite a complicated competition, as the most successful members of the set of 131,072 gradually asserted themselves in successive population censuses. But even this would be a limited kind of heredity, because 131,072, though a large number, is finite. For a life explosion worthy of the name, heredity is needed but so also is indefinite, open-ended variety.

With monensin, we have reached the end of the road, as far as mirror-image heredity is concerned. But left-handedness versus right-handedness is not the only kind of difference that might lend itself to hereditary copying. Julius Rebek and his colleagues at the Massachusetts Institute of Technology are chemists who have taken seriously the challenge of producing self-replicating molecules. The variants they exploit are not mirror images. Rebek and colleagues took two small molecules—the detailed names don't matter, let's just call them A and B. When A and B are mixed in solution, they join up to form a third compound called—you've guessed it—C. Each C molecule acts as a template, or mold. The As and Bs, floating free in solution, find themselves slotting into the mold. One A and one B are jostled into position in the mold, and they thereby find themselves correctly aligned to make a new C, just like the previous one. The Cs don't stick together to form a crystal but split apart. Both Cs are now available as templates to make new Cs, so the population of Cs grows exponentially.

As described so far, the system doesn't exhibit true heredity, but mark the sequel. The B molecule comes in a variety of forms, each of which combines with A to make its own version of the C molecule. So we have C1, C2, C3, and so on. Each of these versions of the C molecule serves as a template for the formation of other Cs of its own type. The population of Cs is therefore heterogeneous. Moreover, the different types of C are not all equally efficient at making daughters. So there is competition between rival versions of C in the population of C molecules. Better yet, "spontaneous mutation" of the C molecule can be induced by ultraviolet radiation. The new mutant type proved to "breed true," producing daughter molecules just like itself. Satisfyingly, the new variant outcompeted the parent type and rapidly took over the test-tube world in which these protocreatures had their being. The A/B/C complex is not the only set of molecules that behaves in this way. There's D, E and F, to name just one comparable triplet. Rebek's group has even been able to make self-replicating hybrids of elements of the A/B/C complex and the D/E/F complex.

The truly self-copying molecules we know in nature—the nucleic acids DNA and RNA—have an altogether richer potential for variation. Whereas a Rebek replicator is a chain with only two links, a DNA molecule is a long chain of indefinite length; each of the hundreds of links in the chain can be any one of four kinds; and when a given stretch of DNA acts as a template for the formation of a new molecule of DNA, each of the four kinds acts as a template for a different particular one of the four. The four units, known as bases, are the compounds adenine, thymine, cytosine and guanine, conventionally referred to as A, T, C and G. A always acts as a template for T, and vice versa. G always acts as a template for C, and vice versa. Any conceivable ordering of A, T, C and G is possible and will be faithfully duplicated. Moreover, since DNA chains are of indefinite length, the range of available variation is effectively infinite. This is a potential recipe for an informational explosion whose reverberations can eventually reach out from the home planet and touch the stars.

The reverberations of our solar system's replicator explosion have been confined to the home planet for most of the four billion years since it happened. Only in the last million years has a nervous system capable of inventing a radio technology arisen. And only in the last few decades has that nervous system actually developed radio technology. Now, an expanding shell of information-rich radio waves is advancing outward from the planet at the speed of light.

I say "information rich" because there were already plenty of radio waves ricocheting around the cosmos. Stars radiate in the radio frequencies as well as in the frequencies we know as visible light. There is even some background hiss left over from the original big bang that baptized time and the universe. But it is not meaningfully patterned: it is not information-rich. A radioastronomer on a planet orbiting Proxima Centauri would detect the same background hiss as our radioastronomers but would also notice an altogether more complicated pattern of radio waves emanating from the direction of the star Sol. This pattern would not be recognized as a mixture of four-year-old television programs, but it *would* be recognized as being altogether more patterned and information-rich than the usual background hiss. The Centaurian radioastronomers would report, amid fanfares of excitement, that the star Sol had exploded in the informational equivalent of a supernova (they'd guess, but might not be sure, that it was actually a planet orbiting Sol).

Replication bombs, as we have seen, follow a slower timecourse than supernovas. Our own replication bomb has taken a few billion years to reach the radio threshold—the moment when a proportion of the information overflows from the parent world and starts to bathe neighboring star systems with pulses of meaning. We can guess that information explosions, if ours is typical, pass a graded series of thresholds. The radio threshold and, before that, the language threshold come rather late in the career of a replication bomb. Before these was what—on this planet, at least—can be called the nerve-cells threshold, and before that there was the many-cells threshold. Threshold number one, the granddaddy of them all, was the replicator threshold, the triggering event that made the whole explosion possible.

What is so important about replicators? How can it be that the chance arising of a molecule with the seemingly innocuous property of serving as a mold for the synthesis of another one just like itself is the trigger of an explosion whose ultimate reverberations may reach out beyond the planets? As we have seen, part of the power of replicators lies in exponential growth. Replicators exhibit exponential growth in a particularly clear form. A simple example is the so-called chain letter. You receive in the mail a postcard on which is written: "Make six copies of this card and send them to six friends within a week. If you do not do this, a spell will be cast upon you and you will die in horrible agony within a month." If you are sensible you will throw it away. But a good percentage of people are not sensible; they are vaguely intrigued, or intimidated by the threat, and send six copies of it to other people. Of these six, perhaps two will be persuaded to send it on to six other people. If, on average, one-third of the people who receive the card obey the instructions written on it, the number of cards in circulation will double every week. In theory, this means that the number of cards in circulation after one year will be 2 to the power 52, or about four thousand trillion. Enough postcards to smother every man, woman and child in the world.

Exponential growth, if not checked by lack of resources, always leads to startlingly large-scale results in a surprisingly short time. In practice, resources are limited, and other factors, too, serve to limit exponential growth. In our hypothetical example, individuals will probably start to balk when the same chain letter comes around to them for the second time. In the competition for resources, variants of the replicator may arise that happen to be more efficient at getting themselves duplicated. These more efficient replicators will tend to replace their less efficient rivals. It is important to understand that none of these replicating entities is consciously interested in getting itself duplicated. But it will just happen that the world becomes filled with replicators that are more efficient.

In the case of the chain letter, being efficient may consist in accumulating a better collection of words on the paper. Instead of the somewhat implausible statement that "if you don't obey the words on the card you will die in horrible agony within a month," the message might change to "Please, I beg of you, to save your soul and mine, don't take the risk: if you have the slightest doubt, obey the instructions and send the letter on to six more people." Such "mutations" can happen again and again, and the result will eventually be a heterogeneous population of messages all in circulation, all descended from the same original ancestor but differing in detailed wording and in the strength and nature of the blandishments they employ. The variants that are more successful will increase in frequency at the expense of less successful rivals. Success is simply synonymous with frequency in circulation. The "St. Jude Letter" is a well-known example of such success; it has traveled around the world a number of times, probably growing in the process. While I was writing this book, I was sent the following version by Dr. Oliver Goodenough, of the University of Vermont, and we wrote a joint paper on it, as a "virus of the mind," for the journal *Nature*:

"WITH LOVE ALL THINGS ARE POSSIBLE"

This paper has been sent to you for luck. The original is in New England. It has been sent around the world 9 times. The Luck has been sent to you. You will receive good luck within 4 days of receiving this letter pending in turn you send it on. This is no joke. You will receive good luck in the mail. Send no money. Send copies to people you think need good luck. Do not send money cause faith has no price. Do not keep this letter. It must leave your hands within 96 hrs. An A.R.P. officer Joe Elliott received $40,000,000. Geo. Welch lost his wife 5 days after this letter. He failed to circulate the letter. However before her death he received $75,000. Please send copies and see what happens after 4 days. The chain comes from Venezuela and was written by Saul Anthony Degnas, a missionary from S.America. Since that copy must tour the world. You must make 20 copies and send them to friends and associates after a few days you will get a surprise. This is love even if you are not superstitious. *Do note* the following: Cantonare Dias received this letter in 1903. He asked his Sec'y to make copies and send them out. A few days later he won a lottery of 20 million dollars. Carl Dobbit, an office employee received the letter and forgot it had to leave his hands within 96 hrs. He lost his job. After finding the letter again he made copies and mailed 20. A few days later he got a better job. Dolan Fairchild received the letter and not believing he threw it away. 9 days later he died. In 1987 the letter was received by a young woman in Calif. It was faded and hardly readable. She promised herself she would retype the letter and send it on but, she put it aside to do later. She was plagued with various problems, including expensive car problems. This letter did not leave her hands in 96 hrs. She finally typed the letter as promised and goe a new car. Remember send no money. Do not ignor this—it works.

St. Jude

This ridiculous document has all the hallmarks of having evolved through a number of mutations. There are numerous errors and infelicities, and there are known to be other versions going around. Several significantly different versions have been sent to me from all around the world since our paper was published in *Nature*. In one of these alternative texts, for instance, the "A.R.P. officer" is an "R.A.F. officer." The St. Jude letter is well known to the United States Postal Service, who report that it goes back before their official records began and exhibits recurrent epidemic outbreaks.

Note that the catalog of alleged good luck enjoyed by compliers and disasters that have befallen refusers cannot have been written in by the victims/beneficiaries themselves. The beneficiaries' alleged good fortune did not strike them until after the letter had left their hands. And the victims did not send the letter out. These stories were presumably just invented—as one might independently have guessed from the implausibility of their content. This brings us to the main respect in which chain letters differ from the natural replicators that initiated the life explosion. Chain letters are originally launched by humans, and the changes in their wording arise in the heads of humans. At the inception of the life explosion there were no minds, no creativity and no intention. There was only chemistry. Nevertheless, once self-replicating chemicals had chanced to arise, there would have been an automatic tendency for more successful variants to increase in frequency at the expense of less successful variants.

As in the case of chain letters, success among chemical replicators is simply synonymous with frequency in circulation. But that is just definition: almost tautology. Success is earned by practical competence, and competence means something concrete and anything but tautological. A successful replicator molecule will be one that, for reasons of detailed chemical technicality, has what it takes to get duplicated. What this means in practice can be almost infinitely variable, even though the nature of the replicators themselves can seem surprisingly uniform.

DNA is so uniform that it consists entirely of variations in sequence of the same four "letters"—A, T, C and G. By comparison, as we have seen in earlier chapters, the means used by DNA sequences to get themselves replicated are bewilderingly variable. They include building more efficient hearts for hippos, springier legs for fleas, aerodynamically more streamlined wings for swifts, more buoyant swim bladders for fish. All the organs and limbs of animals; the roots, leaves and flowers of plants; all eyes and brains and minds, and even fears and hopes, are the tools by which successful DNA sequences lever themselves into the future. The tools themselves are almost infinitely variable, but the recipes for building those tools are, by contrast, ludicrously uniform. Just permutation after permutation of A, T, C and G.

It may not always have been so. We have no evidence that when the information explosion started, the seed code was written in DNA letters. Indeed, the whole DNA/protein-based information technology is so sophisticated—high tech, it has been called by the chemist Graham Cairns-Smith—that you can scarcely imagine it arising by luck, without some other self-replicating system as a forerunner. The forerunner might have been RNA; or it might have been something like Julius Rebek's simple self-replicating molecules; or it might have been something very different: one tantalizing possibility, which I have discussed in detail in *The Blind Watchmaker*, is Cairns-Smith's own suggestion (see his *Seven Clues to the Origin of Life*) of inorganic clay crystals as primordial replicators. We may never know for certain.

What we can do is guess at a general chronology of a life explosion on any planet, anywhere in the universe. The details of what will work must depend on local conditions. The DNA/protein system wouldn't work in a world of chilled liquid ammonia, but perhaps some other system of heredity and embryology would. Anyway, those are just the kinds of specifics I want to ignore, because I want to concentrate on the planet-independent principles of the general recipe. I'll go more systematically now through the list of thresholds that any planetary replication bomb can be expected to pass. Some of these are likely to be genuinely universal. Others may be peculiar to our own planet. It may not always be easy to decide which are likely to be universal and which local, and this question is interesting in its own right.

Threshold 1 is, of course, the Replicator Threshold itself: the arising of some kind of self-copying system in which there is at least a rudimentary form of hereditary variation, with occasional random mistakes in copying. The consequence of Threshold 1's being passed is that the planet comes to contain a mixed population, in which variants compete for resources. Resources will be scarce—or will become scarce when the competition hots up. Some variant replicas will turn out to be relatively successful in competing for scarce resources. Others will be relatively unsuccessful. So now we have a basic form of natural selection.

To begin with, success among rival replicators will be judged purely on the direct properties of the replicators themselves—for example, how well their shape fits a template. But now, after many generations of evolution, we move on to Threshold 2, the Phenotype Threshold. Replicators survive not by virtue of their own properties but by virtue of causal effects on something else, which we call the phenotype. On our planet, phenotypes are easily recognized as those parts of animal and plant bodies that genes can influence. That means pretty well all bits of bodies. Think of phenotypes as levers of power by which successful replicators manipulate their way into the next generation. More generally, phenotypes may be defined as consequences of replicators that influence the replicators' success but are not themselves replicated. For instance, a particular gene in a species of Pacific island snail determines whether the shell coils to the right or to the left. The DNA molecule itself is not right- or left-handed, but its phenotypic consequence is. Left-handed and right-handed shells may not be equally successful at the business of providing the outer protection for snail bodies. Because snail genes ride inside the shells whose shape they help to influence, genes that make successful shells will come to outnumber genes that make unsuccessful shells. Shells, being phenotypes, do not spawn daughter shells. Each shell is made by DNA, and it is DNA that spawns DNA.

DNA sequences influence their phenotypes (like the direction of coiling of shells) via a more or less complicated chain of intermediate events, all subsumed under the general heading of "embryology." On our planet, the first link in the chain is always the synthesis of a protein molecule. Every detail of the protein molecule is precisely specified, via the famous genetic code, by the ordering of the four kinds of letters in the DNA. But these details are very probably of local significance only. More generally, a planet will come to contain replicators whose consequences (phenotypes) have beneficial effects, by whatever means, on the replicators' success at getting copied. Once the Phenotype Threshold is crossed, replicators survive by virtue of proxies, their consequences on the world. On our planet, these consequences are usually confined to the body in which the gene physically sits. But this is not necessarily so. The doctrine of the Extended Phenotype (to which I have devoted a whole book with that title) states that the phenotypic levers of power by which replicators engineer their long-term survival do not have to be limited to the replicators' "own" body. Genes can reach outside particular bodies and influence the world at large, including other bodies.

I don't know how universal the Phenotype Threshold is likely to be. I suspect that it will have been crossed on all those planets where the life explosion has proceeded beyond a very rudimentary stage. And I suspect that the same is true of the next threshold in my list. This is Threshold 3, the Replicator Team Threshold, which may on some planets be crossed before, or at the same time as, the Phenotype Threshold. In early days, replicators are probably autonomous entities bobbing about with rival naked replicators in the headwaters of the genetic river. But it is a feature of our modern DNA/protein information-technology system on Earth that no gene can work in isolation. The chemical world in which a gene does its work is not the unaided chemistry of the external environment. This, to be sure, forms the background, but it is quite a remote background. The immediate and vitally necessary chemical world in which the DNA replicator has its being is a much smaller, more concentrated bag of chemicals—the cell. In a way, it is misleading to call a cell a bag of chemicals, because many cells have an elaborate internal structure of folded membranes on which, in which, and between which vital chemical reactions go on. The chemical microcosm that is the cell is put together by a consortium of hundreds—in advanced cells, hundreds of thousands—of genes. Each gene contributes to the environment, which they all then exploit in order to survive. The genes work in teams. We saw this from a slightly different angle in chapter 1.

The simplest of autonomous DNA-copying systems on our planet are bacterial cells, and they need at least a couple of hundred genes to make the components they need. Cells that are not bacteria are called eukaryotic cells. Our own cells, and those of all animals, plants, fungi and protozoa, are eukaryotic cells. They typically have tens or hundreds of thousands of genes, all working as a team. As we saw in chapter 2, it now seems probable that the eukaryotic cell itself began as a team of half a dozen or so bacterial cells that clubbed together. But this is a higher-order form of teamwork and is not what I am talking about here. I am talking about the fact that all genes do their work in a chemical environment put together by a consortium of genes in the cell.

Once we have grasped the point about genes working in teams, it is obviously tempting to leap to the assumption that Darwinian selection nowadays chooses among rival teams of genes—to assume that selection has moved up to higher levels of organization. Tempting, but in my view wrong at a profound level. It is much more illuminating to say that Darwinian selection still chooses among rival genes, but the genes that are favored are those that prosper *in the presence of the other genes* that are simultaneously being favored in one another's presence. This is the point we met in chapter 1, where we saw that genes sharing the same branch of the digital river tend to become "good companions."

Perhaps the next major threshold to be crossed as a replication bomb gathers momentum on a planet is the Many-Cells Threshold, and I'll call this Threshold 4. Any one cell in our life form, as we saw, is a little local sea of chemicals in which a team of genes bathe. Although it contains the whole team, it is made by a subset of the team. Now, cells themselves multiply by splitting in half, with each one growing to full size again. When this happens, all the members of the team of genes are duplicated. If the two cells do not separate fully but remain attached to one another, large edifices can form, with cells playing the role of bricks. The ability to make many-celled edifices may well be important on other worlds as well as our own. When the Many-Cells Threshold has been crossed, phenotypes can arise whose shapes and functions are appreciated only on a scale hugely greater than the scale of the single cell. An antler or a leaf, an eye's lens or a snail's shell—all these shapes are put together by cells, but the cells are not miniature versions of the large shape. Many-celled organs, in other words, do not grow the way crystals do. On our planet, at least, they grow more like buildings, which are not, after all, the shape of overgrown bricks. A hand has a characteristic shape, but it is not made of hand-shaped cells, as it would be if phenotypes grew like crystals. Again like buildings, many-celled organs acquire their characteristic shapes and sizes because layers of cells (bricks) follow rules about when to stop growing. Cells must also, in some sense, know where they sit in relation to other cells. Liver cells behave as if they know that they are liver cells and know, moreover, whether they are on the edge of a lobe or in the middle. How they do this is a difficult question and a much studied one. The answers are probably local to our planet and I shall not consider them further here. I have already touched upon them in chapter 1. Whatever their details, the methods have been perfected by exactly the same general process as all other improvements in life: the nonrandom survival of successful genes judged by their effects—in this case, effects on cell behavior in relation to neighboring cells.

The next major threshold I want to consider, because I suspect that it, too, is probably of more than local planetary significance, is the High-Speed Information-Processing Threshold. On our planet this Threshold 5 is achieved by a special class of cells called neurons, or nerve cells, and we might locally call it the Nervous System Threshold. However it may be achieved on a planet, it is important, because now action can be taken on a timescale much faster than the one that genes, with their chemical levers of power, can achieve directly. Predators can leap at their dinner and prey can dodge for their lives, using muscular and nervous apparatus that acts and reacts at speeds hugely greater than the embryological origami speeds with which the genes put the apparatus together in the first place. Absolute speeds and reaction times may be very different on other planets. But on any planet an important threshold is crossed when the devices built by replicators start to have reaction times orders of magnitude faster than the embryological machinations of the replicators themselves. Whether the instruments will necessarily resemble the objects that we, on this planet, call neurons and muscle cells is less certain. But on those planets where something equivalent to the Nervous System Threshold is passed, important further consequences are likely to follow and the replication bomb will proceed on its outward journey.

Among these consequences may be large aggregations of data-handling units—"brains"—capable of processing complex patterns of data apprehended by "sense organs" and capable of storing records of them in "memory." A more elaborate and mysterious consequence of crossing the neuron threshold is conscious awareness, and I shall call Threshold 6 the Consciousness Threshold. We don't know how often this has been achieved on our planet. Some philosophers believe that it is crucially bound up with language, which seems to have been achieved once only, by the bipedal ape species *Homo sapiens*. Whether or not consciousness requires language, let us anyway recognize the Language Threshold as a major one,

Threshold 7, which may or may not be crossed on a planet. The details of language, such as whether it is transmitted by sound or some other physical medium, must be relegated to local significance.

Language, from this point of view, is the networking system by which brains (as they are called on this planet) exchange information with sufficient intimacy to allow the development of a cooperative technology. Cooperative technology, beginning with the imitative development of stone tools and proceeding through the ages of metal-smelting, wheeled vehicles, steam power and now electronics, has many of the attributes of an explosion in its own right, and its initiation therefore deserves a title, the Cooperative Technology Threshold, or Threshold 8. Indeed, it is possible that human culture has fostered a genuinely new replication bomb, with a new kind of self-replicating entity—the meme, as I have called it in *The Selfish Gene*—proliferating and Darwinizing in a river of culture. There may be a meme bomb now taking off, in parallel to the gene bomb that earlier set up the brain/culture conditions that made the take-off possible. But that, again, is too big a subject for this chapter. I must return to the main theme of the planetary explosion and note that, once the stage of cooperative technology has been reached, it is quite likely that somewhere along the way the power to make an impact outside the home planet will be achieved. Threshold 9, the Radio Threshold, is passed, and it now becomes possible for external observers to notice that a star system has newly exploded as a replication bomb.

The first inkling external observers will get, as we have seen, is likely to be radio waves spilling outward as a by-product of communications within the home planet. Later, the technological heirs of the replication bomb may themselves turn their deliberate attention outward to the stars. Our own halting steps in that direction have included the beaming out into space of messages specifically tailored for alien intelligences. How can you tailor messages for intelligences of whose nature you have no conception? Obviously it is difficult, and quite possibly our efforts have been misconceived.

Most attention has been given to persuading alien observers that we exist at all, rather than sending them messages with substantial content. This task is the same as that faced by my hypothetical Professor Crickson in chapter 1. He rendered the prime numbers into the DNA code, and a parallel policy using radio would be a sensible way to flag our presence to other worlds. Music might seem a better advertisement for our species, and even if the audience lacked ears they might appreciate it in their own way. The famous scientist and writer Lewis Thomas suggested that we broadcast Bach, all of Bach and nothing but Bach, although he feared it might be taken as boasting. But, equally, music might be mistaken, by a sufficiently alien mind, for the rhythmic emanations of a pulsar. Pulsars are stars that give off rhythmic pulses of radio waves at intervals of a few seconds or less. When they were first discovered, by a group of Cambridge radioastronomers in 1967, there was momentary excitement as people wondered whether the signals might be a message from space. But it was soon realized that a more parsimonious explanation was that a small star was rotating extremely fast and sweeping a beam of radio waves around like a lighthouse. To date, no authenticated communications from outside our planet have ever been received.

After radio waves, the only further step we have imagined in the outward progress of our own explosion is physical space travel itself: Threshold 10, the Space Travel Threshold. Science-fiction writers have dreamed of the interstellar proliferation of daughter colonies of humans, or their robotic creations. These daughter colonies could be seen as seedings, or infections, of new pockets of self-replicating information—pockets that may subsequently themselves expand explosively outward again, in satellite replication bombs, broadcasting both genes and memes. If this vision is ever realized, it is perhaps not too irreverent to imagine some future Christopher Marlowe reverting to the imagery of the digital river: "See, see, where life's flood streams in the firmament!"

We have so far scarcely taken the first step outward. We have been to the moon but, magnificent as this achievement is, the moon, though no calabash, is so local as scarcely to count as traveling, from the point of view of the aliens with whom we might eventually communicate. We have sent a handful of unmanned capsules into deep space, on trajectories that have no visualizable ending. One of these, as a result of inspiration from the visionary American astronomer Carl Sagan, carries a message designed to be deciphered by any alien intelligence who might chance upon it. The message is adorned with a picture of the species that created it, the image of a naked man and woman.

This might seem to bring us full circle, to the ancestral myths with which we began. But this couple is not Adam and Eve, and the message engraved beneath their graceful forms is an altogether more worthy testament to our life explosion than anything in Genesis. In what is designed to be a universally understandable iconic language, the plaque records its own genesis in the third planet of a star whose coordinates in the galaxy are precisely recorded. Our credentials are further established by some iconic representations of fundamental principles of chemistry and mathematics. If the capsule is ever picked up by intelligent beings, they will credit the civilization that produced it with something more than primitive tribal superstition. Across the gulf of space, they will know that there existed, long ago, another life explosion that culminated in a civilization that would have been worth talking to.

Alas, this capsule's chance of passing within a parsec of another replication bomb is forlornly small. Some commentators see its value as an inspirational one for the population back home. A statue of a naked man and woman, hands raised in a gesture of peace, deliberately sent on an eternal outward journey among the stars, the first exported fruit of the knowledge of our own life explosion—surely the contemplation of this might have some beneficial effects upon our normally parochial little consciousnesses; some echo of the poetic impact of Newton's statue in Trinity College, Cambridge, upon the admittedly giant consciousness of William Wordsworth:

> And from my pillow, looking forth by light
> Of moon or favouring stars, I could behold
> The antechapel where the statue stood
> Of Newton with his prism and silent face,
> The marble index of a mind for ever
> Voyaging through strange seas of Thought, alone.

BIBLIOGRAPHY AND FURTHER READING

With a few exceptions I have limited this list to readily accessible books rather than technical works that can be found only in university libraries.

Bodmer, Walter, and Robin McKie, *The Book of Man: The Human Genome Project and the Quest to Discover Our Genetic Heritage* (New York: Scribners, 1995).

Bonner, John Tyler, *Life Cycles: Reflections of an Evolutionary Biologist* (Princeton: Princeton University Press, 1993).

Cain, Arthur J., *Animal Species and Their Evolution* (New York: Harper Torchbooks, 1960).

Cairns-Smith, A. Graham, *Seven Clues to the Origin of Life* (Cambridge: Cambridge University Press, 1985).

Cherfas, Jeremy, and John Gribbin, *The Redundant Male: Is Sex Irrelevant in the Modern World?* (New York: Pantheon, 1984).

Clarke, Arthur C., *Profiles of the Future: An Inquiry into the Limits of the Possible* (New York: Holt, Rinehart & Winston, 1984).

Crick, Francis, *What Mad Pursuit: A Personal View of Scientific Discovery* (New York: Basic Books, 1988).

Cronin, Helena, *The Ant and the Peacock: Altruism and Sexual Selection from Darwin to Today* (New York: Cambridge University Press, 1991).

Darwin, Charles, *The Origin of Species* (New York: Penguin, 1985).

———, *The Various Contrivances by Which Orchids are Fertilised by Insects* (London: John Murray, 1882).

Dawkins, Richard, *The Extended Phenotype* (New York: Oxford University Press, 1989).

———, *The Blind Watchmaker* (New York: W.W. Norton, 1986).

———, *The Selfish Gene,* new ed. (New York: Oxford University Press, 1989).

Dennett, Daniel C., *Darwin's Dangerous Idea* (New York: Simon & Schuster, 1995).

Drexler, K. Eric, *Engines of Creation* (Garden City, N.Y.: Anchor Press/Doubleday, 1986).

Durant, John R., ed. *Human Origins* (Oxford: Oxford University Press, 1989).

Fabre, Jean-Henri, *Insects,* David Black, ed. (New York: Scribners, 1979).

Fisher, Ronald A., *The Genetical Theory of Natural Selection,* 2d. rev. ed. (New York: Dover, 1958).

Frisch, Karl von, *The Dance Language and Orientation of Bees,* Leigh E. Chadwick, trans. (Cambridge: Harvard University Press, 1967).

Gould, James L., and Carol G. Gould, *The Honey Bee* (New York: Scientific American Library, 1988).

Gould, Stephen J., *Wonderful Life: The Burgess Shale and the Nature of History* (New York: W.W. Norton, 1989).

Gribbin, John, and Jeremy Cherfas, *The Monkey Puzzle: Reshaping the Evolutionary Tree* (New York: Pantheon, 1982).

Hein, Piet, with Jens Arup, *Grooks* (Garden City, N.Y.: Doubleday, 1969).

Hippel, Arndt von, *Human Evolutionary Biology* (Anchorage: Stone Age Press, 1994).

Humphrey, Nicholas K., *Consciousness Regained* (Oxford: Oxford University Press, 1983).

Jones, Steve, Robert Martin and David Pilbeam, eds., *The Cambridge Encyclopedia of Human Evolution* (New York: Cambridge University Press, 1992).

Kingdon, Jonathan, *Self-made Man: Human Evolution from Eden to Extinction?* (New York: Wiley, 1993).

Macdonald, Ken C., and Bruce P. Luyendyk, "The Crest of the East Pacific Rise," *Scientific American*, May 1981, pp. 100–116.

Manning, Aubrey, and Marian S. Dawkins, *An Introduction to Animal Behaviour,* 4th ed. (New York: Cambridge University Press, 1992).

Margulis, Lynn, and Dorion Sagan, *Microcosmos: Four Billion Years of Microbial Evolution* (New York: Simon & Schuster, 1986).

Maynard Smith, John, *The Theory of Evolution* (Cambridge: Cambridge University Press, 1993).

Meeuse, Bastiaan, and Sean Morris, *The Sex Life of Plants* (London: Faber & Faber, 1984).

Monod, Jacques, *Chance and Necessity: An Essay on the Natural Philosophy of Modern Biology,* Austryn Wainhouse, trans. (New York: Knopf, 1971).

Nesse, Randolph, and George C. Williams, *Why We Get Sick: The New Theory of Darwinian Medicine* (New York: Random House, 1995).

Nilsson, Daniel E., and Susanne Pelger, "A Pessimistic Estimate of the Time Required for an Eye to Evolve," *Proceedings of the Royal Society of London, B* (1994).

Owen, Denis, *Camouflage and Mimicry* (Chicago: University of Chicago Press, 1982).

Pinker, Steven. *The Language Instinct: The New Science of Language and the Mind* (New York: Morrow, 1994).

Ridley, Mark, *Evolution* (Boston: Blackwell Scientific, 1993).

Ridley, Matt., *The Red Queen: Sex and the Evolution of Human Nature* (New York: Macmillan, 1994).

Sagan, Carl, *Cosmos* (New York: Random House, 1980).

———, and Ann Druyan, *Shadows of Forgotten Ancestors* (New York: Random House, 1992).

Tinbergen, Niko, *The Herring Gull's World* (New York: Harper & Row, 1960).

———, *Curious Naturalists* (London: Penguin, 1974).

Trivers, Robert, *Social Evolution* (Menlo Park, Calif.: Benjamin-Cummings, 1985).

Watson, James D., *The Double Helix: A Personal Account of the Discovery of the Structure of DNA* (New York: Atheneum, 1968).

Weiner, Jonathan, *The Beak of the Finch: A Story of Evolution in Our Time* (New York: Knopf, 1994).

Wickler, Wolfgang, *Mimicry in Plants and Animals,* R. D. Martin, trans. (New York: McGraw-Hill, 1968).

Williams, George C., *Natural Selection: Domains, Levels, and Challenges* (New York: Oxford University Press, 1992).

Wilson, Edward O., *The Diversity of Life* (Cambridge: Harvard University Press, 1992).

Wolpert, Lewis, *The Triumph of the Embryo* (New York: Oxford University Press, 1992).

INDEX